ADVANCES
IN DISTRIBUTED
SENSOR INTEGRATION

ADVANCES IN DISTRIBUTED SENSOR INTEGRATION

Application and Theory

Series Editor

Mohammad Jamshidi   University of New Mexico

Authors

S.S.Iyengar   Louisiana State University
L. Prasad   Los Alamos National Laboratory
Hla Min   Louisiana State University

Prentice Hall PTR
Upper Saddle River, New Jersey 07458

Library of Congress Cataloging-in Publication Data

Iyengar, S.S. (Sundararaja S.)
      Advances in distributed sensor integration : theory and
application / S.S. Iyengar, L. Prasad, H. Min.
          p.   cm.
      Includes bibliographical references and index.
      ISBN 0-13-360033-5
      1.Signal processing. 2. Multisensor data fusion--Mathematics.
I. Prasad, L. (Laksham) II. Min, H. (Hla) III. Title.
TK5105.9.I94   1995   95-3383
621.382'2--dc20   CIP

Editorial/production supervision: *Harriet Tellem*
Cover design: *Aren Graphics*
Manufacturing manager: *Alexis R. Heydt*
Acquisitions editor: *Bernard Goodwin*
Editorial assistant: *Diane Spina*

ISBN 0-13-360033-5

To Our Families

Contents

Preface

Sensor processing is a central and important problem in Aerospace/Defense, Automation, Medical Imaging, and Robotics, to name a few areas. A surveillance system used in aerospace and defense is an example of a *sensor processing system*. It uses devices such as infrared sensors, microwave radars and laser radars, which are capable of detecting and tracking flying objects in their observation space.

A sensor processing system may employ intelligent and disparate sensors that are distributed logically, spatially, and even geographically. It is then referred to as a *Distributed Sensor Network* (DSN).

The sensors may measure scalar values (e.g., temperature) or vector values (e.g., position in three-dimensional space). The measurements are generally a function of time and/or space. Due to differences in operating environments or other factors, such as aging and commnunication delays, the measurements might be susceptible to errors; some sensor measurements may appear contradictory. Combining the numerous sensor measurements to minimize the uncertainty and to improve the reliability and fault tolerance is a central theme of the book.

This book attempts to complement existing literature. Distributed sensor processing is interdisciplinary in nature; it shares issues with signal and image processing, distributed systems, computer science, data communications, and intelligent systems. We will give sufficient background material for experts in other disciplines to appreciate the formal and computational properties of distributed sensor processing, in general, and the algorithms, data structures, architectures, and computational complexity of sensor integration in a distributed environment. We will also discuss simple distributed detection problems and their computational complexities.

Our research papers form the core of this book. They have been funded in part by Office of Naval Research grant ONR N0014-91J-1306 and in part by the Innovative Science and Technology (IST) Program of the SDIO monitored by the Office of Naval Research under grant ONR N00014-85K-0611 on Distributed Sensor Networks. The papers have been revised both for their content and for style. They have been enhanced with examples, explanatory notes, and problems to make the book self-contained for use in senior-level undergraduate or first-year graduate courses.

We discuss the development of fault-tolerant algorithms for integrating (a) scalar-valued sensor readings, (b) vector-valued sensor readings, and (c) sensors in a given DSN architecture.

We use a top-down design of the algorithms, give a pseudocode description, and then perform a worst-case analysis of algorithms. Proof of correctness and/or optimality of some algorithms is given where appropriate.

We also cover algorithms that are related to sensor processing. For example, the diameter of a network is an important parameter and can be critical in real-time applications. Distributed algorithms for finding the diameter of an asynchronous tree network are given.

Distributed Detection Problems are interesting from the computational complexity viewpoint. While some simple detection problems have polynomial solutions, some versions are computationally intractable. After investigating the factors that make them intractable, we will relax the optimality constraint and then develop reasonably good approximate algorithms for such problems.

We will characterize the *sensor integration problem*. We will also give a computational structure for a distributed sensor network.

In the quest for efficient computer architectures for sensor integration, we first review two architectures that have been proposed: committee organization and hierarchical organization. The advantages and limitations lead us to propose hybrid organizations: a *flat tree* architecture and also a *multi-level binary de Bruijn* (MBD) network.

Acknowledgements

This book evolved from the research papers on sensor integration in a distributed environment. We are indebted to the Office of Naval Research, National Science Foundation, and Louisiana Education Quality Support Foundation(LEQSF)-Board of Regents for funding the research projects.

This work is made possible by the contributions of our collaborators which include Dr. R. N. Madan (Office of Naval Research), Professor R. L. Kashyap (Purdue University), Dr. N. S. V. Rao (Oak Ridge National Laboratory), Dr. D. N. Jayasimha (Ohio State University, Columbus), Dr. M. Sharma (IBM, Austin), and Ramana Rao, Daryl Thomas, Deepak Nadig, and Richard R. Brooks (all of Louisiana State University).

We wish to thank the following journals in which our work originally appeared for the permission to use excerpts from the articles.

- *IEEE Transactions on Computers*

- *IEEE Transactions on Signal Processing*

- *IEEE Transactions on Systems, Man and Cybernetics*

- *Journal of Computer Science and Information*

- *Journal of Computers and Electrical Engineering*

- *Physical Review E*

- *Journal of Theoretical Computer Science*

- *Information Processing Letters*

We also wish to thank Yuyan Wu, Ramana Rao, Richard R. Brooks, John Zachary, J.R. Maheshkumar, and Vijayashankar V. for their contributions. Dr. R. N. Madan, Professor R. L. Kashyap, Dr. N. S. V. Rao, Dr. F. Bastani, and Richard R. Brooks reviewed preliminary versions of this book. We also thank the anonymous reviewers, whose comments have contributed to a more coherent presentation of our work.

Chapter 1

Introduction and Mathematical Preliminaries

Given the tremendous growth in the field of signal processing and sensor integration where technologies change frequently, we feel that another book with an emphasis on technical details would be futile. Consequently, our focus has been to provide insight into the algorithmic issues and needs for building a unisensor or multisensor system. For each component subsystem, based on a comprehensive evaluation of state-of-the-art technology, we identify a viable algorithmic kernel for provision of the envisioned ability of the subsystem. Concomitant to the algorithmic analysis is our constant focus on computational complexity and real-time performance. Our motivation for the latter emphasis stems from our conviction that, despite the technological and engineering strides in subsystem development and availability, provision of an integrated intelligent sensor capability needed for unstructured applications is a long way off. With this in mind, we have constantly reinforced the compromise between tractability and capability.

Sensor integration is central to industrial, medical, space, and military applications, to name a few areas. This interdisciplinary nature leads us to expect that this book will be read by professionals with varying backgrounds and expertise. It is with this audience in mind that this chapter aims to provide some background material on signal processing, discrete mathematics, and computer systems. Readers may want to casually peruse the material and refer to details throughout the study of this book.

1.1 Fundamental Concepts Of Signal Processing

Our world today is bathed in a sea of signals. Intuitively, we are aware of the concept of a signal by our experiences with speech and music. Images and light are also in our realm of signal experience. As humans, we are great signal processors, being able to process one dimensional (speech, music) and two-dimensional (images, light) signals with unconscious effort. Yet we find it quite difficult to build machines and software to process signals at least as efficiently as a human.

An example of a signal processor is the omnipresent telephone. Acoustic signals originating from the speaker are converted to electrical signals (and in some cases optical signals) and transmitted to a remote location, where they are converted back to acoustic signals and perceived by a listener.

Associated with signals is error, sometimes called *noise*. Background noise that invades every signal is called white, or Gaussian, noise. Random surges or suppressions called interference may also be present in signals.

Signal processing involves many tasks, such as signal analysis, diagnostics, coding, error correction, signal compression, quantization, and transmission. Signals have an *amplitude*, which describes the intensity of the signal or the maximum displacement from the equilibrium position, and a *frequency*, which describes the speed of the signal or the inverse of the signals period. Signals may be characterized as spatial or time series.

Signal synthesis and analysis employ several transforms, such as the Fourier transform, the Gabor transform (or Window Fourier Transform), the z-transform, the Laplace transform, and several wavelet transforms. As may be inferred, transform theory alone is a rich and diverse field of signal processing.

1.2   Algorithms

An algorithm is defined as a step by step procedure for solving a computational problem in a finite number of steps. The logical instructions for each step should be precise and the algorithm should terminate after a finite number of steps. Probably, the most famous algorithm is Euclid's algorithm for determining the greatest common divisor of any two integers.

Algorithms can be classified in various ways, as follows:

1. Sequential, parallel, or distributed
2. Deterministic, nondeterministic or random
3. Heuristic or nonheuristic
4. Stochastic

An algorithm where the different steps are executed strictly in sequence is called a sequential algorithm. An algorithm, where several steps can be simultaneously executed in parallel is called a parallel algorithm. A distributed algorithm is one where the many steps of the algorithm are executed on different machines and in an asynchronous manner.

If at each step of an algorithm, based on whether a condition is true or false a definite action can be taken, then it is a deterministic algorithm. If, the algorithm simultaneously explores a number of alternatives to arrive at the correct action to be taken, it is called nondeterministic. If, after each step of execution, the next step is decided by a random process we call it a random algorithm.

A heuristic algorithm makes use of some *knowledge* based on previous experience to arrive at a reasonably optimal solution to the problem. A nonheuristic algorithm attempts to arrive at an optimal solution based on a definite mathematical model of the system.

A stochastic algorithm is an algorithm which represents a system having a number of random components. (e.g., fusion of two metals).

For some algorithms, it is possible to give a proof of correctness.

Algorithms are often classified by their computational complexity, for example time complexity, space complexity, and message complexity. Sequential algorithms are considered efficient if their (time) complexity is polynomial in the size of the input. If no known polynomial-time solutions are known for a problem, the problem is said to be intractable. For intractable problems, approximation algorithms are used.

Neural net algorithms and evolutionary algorithms (including genetic algorithms) are some of the nontraditional algorithms that are in use.

1.3   Rudiments of Set Theory

Concepts of set theory form the foundation for probability, graph theory, and other branches of discrete mathematics. Although most people have a basic understanding of sets from their experiences as children, the theory of sets is deep and rich with still unsolved problems. We will provide a basic framework of set theory to be used in understanding other concepts in discrete mathematics.

1.3.1   Definitions

A *set* may be defined as a collection of elements that are distinct from one another. Some trivial examples of sets are

$$A = \{\text{penny, nickel, dime, quarter, half-dollar, dollar}\} = \text{set of U.S. currency}$$
$$B = \{\text{Monday, Tuesday, Wednesday, Thursday, Friday, Saturday, Sunday}\}$$
$$= \text{set of the days of the week}$$
$$C = \{x : x = 2i,\ i = 1, 2, \dots\} = \text{set of positive even integers}$$

To say that x is an element of S, we write $x \in S$. $x \notin S$ means x is not an element of S.

A set may be finite (example A and B) or infinite (example C). Many times, it is convenient to talk about the size of a set. The number of elements in a set is called the *cardinality* of the set and is denoted $|s|$. From the examples above, $|A| = 6$, $|B| = 7$, and $|C| = \infty$. There is a special set, called the *empty set* or *null set*, with cardinality 0.

In some sets, elements are not distinct. Take, for instance, the set of names of people in a room. Clearly, some people may have the same name, for example, "John" is common. Thus, a name set may be

$$\{\text{John, Lori, George, Darlene, Sue, John}\}.$$

where the name "John" is repeated twice. Thus, a collection of elements is called a *multiset*. Notice the property of distinctness between elements is not part of this definition.

In many applications, it is useful to know if a collection of elements belongs to a set. If so, this collection is known as a *subset*. If S is a subset of T, we denote this symbolically as $S \subseteq T$. In more rigorous terms, every element of S is an element of T. As an example, the set of even integers is a subset of the set of integers. If T contains at least one element that is not an element of S, then S is said to be a proper subset of T, denoted $S \subset T$. It is easy to prove that the even integers are a proper subset of the set of integers. Another easy axiom to show is that if $S \subseteq T$ and $T \subseteq S$, then $S = T$.

In the context of any set problem, there is a *universal set*. All sets are a subset of the universal set. In our discussion, the universal set is denoted by $\mathbf{U}$.

Up to this point, it has been implied that the order of appearance of elements in a set is not important, that is $\{1, 2, 3, 4, 5\} = \{2, 5, 4, 3, 1\}$. However, in many sensor applications, it is useful to keep elements in an ordered fashion. Sets for which the order of the elements matter are called *ordered sets*. In some mathematical contexts, ordered sets are called *permutations*. For example, consider the previously described set. $S = \{1, 2, 3, 4, 5\}$ is a set of five integers. Some permutations of S are $\{1, 3, 2, 4, 5\}$, $\{1, 4, 2, 3, 5\}$, and $\{2, 5, 4, 3, 1\}$. In general, if we are given a set of n integers $(1, 2, ..., n)$, there are $n!$ unique permutations, or ordered sets.

1.3.2　Set Operations

We wish to focus our discussion on the basic operations and properties of sets. In the discussion that follows, we will let S and T be two arbitrary and unordered sets.

The *union* of a two sets is defined as $S \cup T = \{x : x \in S \;\; or \;\; x \in T\}$.

The *intersection* of two sets is defined as $S \cap T = \{x : x \in S \; and \; x \in T\}$.

The *complement* of a set S is defined as $\neg S = \{x : x \notin S\}$.

Another definition for the set complement is $\neg S = U - S$.

The key distinction between intersection and union is the difference between the Boolean operators "or" and "and". The underlying Boolean operator in the set complement is "not". As might be guessed, there is a fundamental relationship between Boolean algebra and set operations.

Although these operations are elementary, we wish to make them concrete with an example.

Example
$$A = \{1, 4, 6, 7, 9\}$$
$$B = \{1, 2, 5, 6, 8\}$$
$$C = \{x: x \text{ is an even integer}\}$$
We assume that $\mathbf{U} = \{$set of integers$\}$. Then,
$$A \cup B = \{1, 4, 6, 7, 9\} \cup \{1, 2, 5, 6, 8\} = \{1, 2, 4, 5, 6, 7, 8, 9\},$$
$$A \cap B = \{1, 4, 6, 7, 9\} \cap \{1, 2, 5, 6, 8\} = \{1, 6\},$$

$$A \cap B \cap C = \{1, 4, 6, 7, 9\} \cap \{1, 2, 5, 6, 8\} \cap \quad \{x: x \text{ is an even integer}\}$$
$$= \{x: x \text{ is an even integer}\} \qquad = C, \text{ and}$$
$$\neg C \qquad = \neg \{x: x \text{ is an even integer}\} \quad = \{y: y \text{ is an odd integer}\}.$$

Sets may be represented in the usual set notation or in terms of pictures called *Venn diagrams*. Venn diagrams represent the universal set as a square and subsets of the universal set as circles or ellipses inside the square as in Figure 1.1. Shaded areas denote set inclusion, that is, elements of sets.

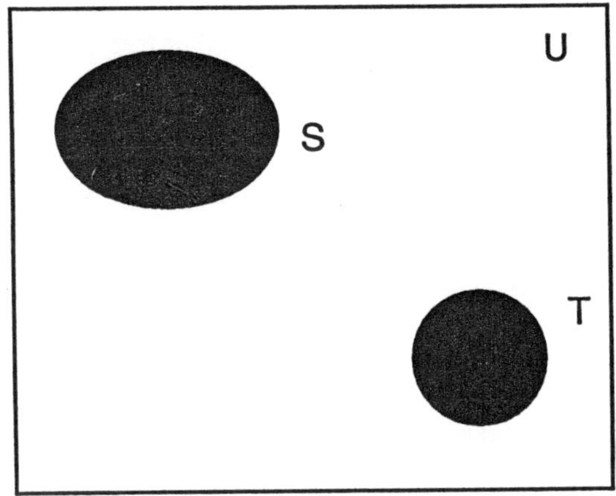

Figure 1.1 $S \cup T = \emptyset$.

If $S \cap T = \emptyset$, then S and T are said to be *disjoint*. Another common term is to say that S and T are *mutually exclusive*. Simply, S and T have no elements in common. In the context of the universal set $\mathbf{U}$, S and T are said to be *collectively exhaustive* if $S \cup T = \mathbf{U}$. If S and T are mutually exclusive and collectively exhaustive, $S = \neg T$.

The *Cartesian product* of sets S and T is defined as $S \times T = \{(s,t) : s \in S \text{ and } t \in T\}$, and $|S \times T| = |S| \times |T|$. $S \times T$ defines points in two-dimensional space. We may also consider the Cartesian product of more than two sets, $|S_1 \times S_2 \times \cdots \times S_n| = \{(s_1, s_2, ..., s_n) : s_i \in S_i, 1 \leq i \leq n\}$. Thus, we have a set of points in n-dimensional space.

There is a special set of S called the *power set* of S, denoted 2^S. This set is the union of all subsets of S, including the proper subsets.

To illustrate these concepts, we have the following examples.

Let

$$A \quad = \{a, c, q, w, z\}$$

$$B \quad = \{1, 4, 5, 7, 9\}$$

Then

$$A \times B = \{(a, 1), (a, 4), (a, 5), (a, 7), (a, 9), (c, 1), (c, 4), (c, 5), (c, 7), (c, 9),$$
$$(q, 1), (q, 4), (q, 5), (q, 7), (q, 9), (w, 1), (w, 4), (w, 5), (w, 7), (w, 9),$$
$$(z, 1), (z, 4), (z, 5), (z, 7), (z, 9)\}.$$

Let $C = \{\text{ultrasonic, infrared, laser}\}$ be the sensor suite for a laboratory's robot. The power set of the sensor suite is

$$2^C = \{ \emptyset, (\text{ultrasonic}), (\text{infrared}), (\text{laser}), (\text{ultrasonic, infrared}),$$
$$(\text{ultrasonic, laser}), (\text{infrared, laser}), (\text{ultrasonic, infrared, laser})\}.$$

In some remote environments, it is useful to know exactly what sensors are transmitting data. Here we see that, given three sensors, there are eight possibilities of sensor activity.

1.3.3 Representation of Finite Sets

How can we represent a finite set, say A, where $|A| = n$? There are four possibilities:

1. A is an arbitrary set.

2. A is a linearly ordered set.

3. A is a set with a range of integer values.

4. A is a consecutively by numbered set.

By an arbitrary set, we mean that no ordering (or structure) is known about its members. To implement such a set, we can use a linked list.

A set A is set to be a linearly ordered set if it has a linear order associated with it. For any two members $a, b \in A$, either $a < b$ or $b < a$. Suppose the members of A are names. Then the lexicographic order of the names can be used as a linear order. To implement such a set, we can build a binary search tree in $\Theta(n \log n)$ time. Suppose each member must come from a range of integer values, say 1 to M. We can sort A in $\Theta(n + M)$ using bucket sort.

Assume $A = \{1, 2, \ldots, n\}$. This is more restricted than the previous case, where there can be gaps in the integer values. We can sort set A in $\Theta(n)$ using a Boolean array.

1.3.4 Comparing Two Subsets for Equality

Let A, $|A| = p$ and B, $|B| = q$ both be subsets of X, the universe of discourse. Let $|X| = n$. How fast can we find out if $A = B$? As discussed in the previous subsection, there are four cases.

Suppose the sets A and B are arbitrary sets. We simply check if a member of one set is also a member of the other. For the worst case, we need to make pq comparisons. The complexity for this straightforward algorithm is $O(|A| \times |B|)$.

Suppose there is a linear order associated with the sets A and B. Build a binary tree using the smaller set (say A). This requires $O(p \log p)$ time. For each element in the larger set (say B), check for membership of other set (say A) in the binary tree. This takes $O(\log p)$ per element and $(O(q \log p)$ for q elements. The total time taken is $O((p + q) \log p)$, assuming A is the smaller set. The complexity, in the general case, is $O[(|A| + |B|) \log(\min(|A|, |B|))]$.

Suppose the elements of A and B fall in the range 1 to M, where M is a specified integer. Use an array C of size M to map all elements of both sets. Check if each element of C has 0 or 2 elements mapped to it. The complexity is $O(|A| + |B| + M)$.

Assume X is a consecutively numbered set. A Boolean array of size n can be used to determine the membership of a set. Use a Boolean array of size n. The complexity is $O(n)$.

1.4   Relations And Functions

In the previous section we gave a definition of an ordered set. In this section, we will expand on ordered sets to develop the idea of relations and functions, which are fundamental concepts in all of mathematics.

1.4.1   Definitions of Relations

Let A and B be sets and $C = A \times B$. Then, $R \subseteq C$ is called a *binary relation* on A to B. If $(a, b) \in R$, then a is related to b, or $a \ R \ b$ [46]. This definition may be generalized to k relations. If $A_1, A_2, \cdots, A_k$ are sets and $C_k = A_1 \times A_2 \times \cdots \times A_k$, then $R_k \subseteq C_k$ is a k-ary relation. The elements of R_k are called *k-tuples* and are ordered subsets.

1.4.2   Binary Relations

Binary relations arise naturally in many contexts. For this reason, this section is devoted to a discussion of binary relation representations and binary relation properties.

Given a binary relation on an set, we have has several choices of a representation scheme. If the relation set is small, then it is common to use a set theoretic notation or list the elements of the relation. For example, $S = \{(i, j) : i \in \mathbf{Z}^+ \setminus \{0\}$ and $j = \log i\}$ and $T = \{(1, 2), (2, 4), (3, 6), (4, 8)\}$ are binary relations. Another representation strategy is to give a relation matrix. Let R_n be a relation on a set of n elements. Define an 4×4 matrix as $M_n = [m(i, j)]$, where

$$m(i, j) = \begin{cases} 1, & (r_i, r_j) \in R_n \\ 0, & \text{otherwise} \end{cases}$$

If there is a 1 in the ith row and j^{th} column, then there exists a relation between r_i and r_j in R_n. For example, let $R = \{(r_1, r_3), (r_2, r_3), (r_2, r_4), (r_3, r_1), (r_3, r_4), (r_4, r_1)\}$ be a relation. Then the relation matrix for R is shown in Table 1.1.

R	r_1	r_2	r_3	r_4
r_1	0	0	1	0
r_2	0	0	1	1
r_3	1	0	0	1
r_4	1	0	0	0

Table 1.1 Relation Matrix

Another common representation for a relation is to represent R as a directed graph. Graphs are discussed in more depth in section 1.5, and here we simply give a graph for the previous example(see Figure 1.2).

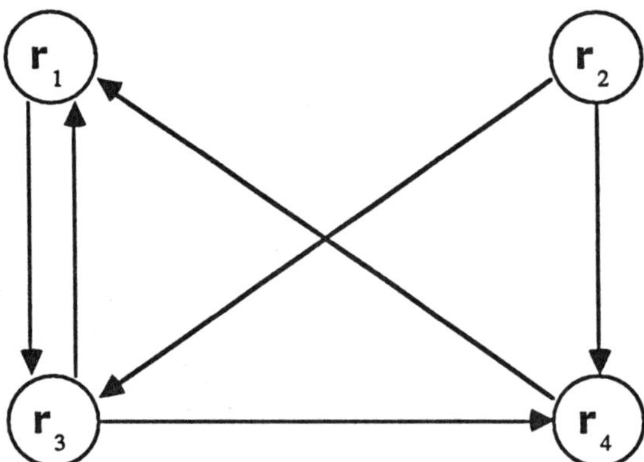

Figure 1.2 Graph.

Binary relations have a set of properties associated with them. Let R be a binary relation on a set S. If $s\,R\,s$ for all $s \in S$, then R is said to be *reflexive*. If $s\,\neg R s$ for all $s \in S$, then R is *irreflexive*. If $s\,R\,t$ implies $t\,R\,s$ for all $s,t \in S$, then R is *symmetric*, and if the implication does not follow, then R is *asymmetric*. If, however, $s\,R\,t$ and $t\,R\,s$ implies that $s = t$, then R is said to be *antisymmetric*. Finally, if $s,t,u \in S$ and $s\,R\,t$ and $t\,R\,u$ imply that $s\,R\,u$, then R is a *transitive* relation.

1.4.3   Equivalence Relations and Partitions

The binary relation properties are used to define a very important class of relations called equivalence relations. *Equivalence relations* are relations that are reflexive, symmetric, and transitive. Equality is the best known equivalence relation, while modulo arithmetic is another important equivalence relation.

Definition: Let R be an equivalence relation on a set S and $s \in S$. Then the set $[s] = \{t \in S : s\ R\ t\}$ is called the *equivalence class* of s. Equivalence classes have a profound property. If $[s]$ and $[t]$ are equivalence classes of an equivalence relation R on a set S, then either $[s] = [t]$ or $[s] \cap [t] = \emptyset$.

Definition: Let S be a set of elements. Then a *partition* of S is a collection of nonempty subsets $\{W_i\}$ of S such that $W_i \cap W_j = \emptyset$ if $i \neq j$. Also, $\cup\{W_i\} = S$.

Intuitively, there is some relationship between partitions and equivalence classes. The following proposition asserts that equivalence classes and partitions are equivalent.

Proposition: If R is an equivalence relation on a set S, then the collection of equivalence classes is the same as a partition of S.

Example: Consider the operation of congruence modulo 2. This equivalence relation on the set of integers $\mathbf{Z}$ defines a partition of the integers into even and odd numbers.

1.4.4   Definitions of Functions

Functions are a natural extension of relations. Assume that S and T are two sets. A *function* of S to T is a mapping by which each element of S is mapped onto exactly one element of T. A partial function of S to T is a mapping by which each element of S is mapped onto at least one element of T. Thus, a partial function is not as strict as a function. We will concern ourselves with functions. If f is a function from S to T, it is denoted as $f : S \mapsto T$. S is called the *domain* of f and T is called the *range* of f. If $s \in S$, $t \in T$, and f maps s onto t, then this is usually denoted as $f(s) = t$, where t is called the image of s. If $t = s$ for all $s \in S$, then f is called an identity function, denoted as I_S. In some functions, s may be a vector, $(s_1, s_2, ..., s_n)$. In this case, f is called a *multivariable function* of dimension n.

1.4.5   Properties of Functions

Let $f : S \mapsto T$ be a function. f is said to be *injective*, or one to one, if for every $s_1, s_2 \in S$ such that $s_1 \neq s_2$ then $f(s_1) \neq f(s_2)$. This is to say that each element in S maps to at most one element in T. f is *surjective*, or onto, if for every $t \in T$ there exists an $s \in S$ such that $f(s) = t$. This property means that each element in S maps to at least one element in T. If a function is both surjective and injective, it is called a bijective function, and each element in S maps to exactly one element in T.

There exists an operation between functions called *functional composition*. Let $f_1 : U \mapsto T$ and $f_2 : S \mapsto U$. Then the functional composition of f_1 and f_2 is $f_1 \bullet f_2$

$= f_1(f_2(t))$. We can alternatively write $f_1 \bullet f_2 : S \mapsto T$, and it is easy to see that without stating any conditions f_2 cannot be a partial function. By example, one may check that the functional composition operator $\bullet$ is not necessarily commutative. Let f_1 be defined by $2x^3 \colon \mathbf{R} \mapsto \mathbf{R}$ and f_2 be defined by $3x \colon \mathbf{R} \bullet \mathbf{R}$. Then $f_1 \bullet f_2 = 2x^3 \bullet 3x = 54x^3$, which is not equal to $f_2 \bullet f_1 = 3x \bullet 2x^3 = 6x^3$.

1.4.6 Inverse Functions

In many applications, sensor data are obtained that describe some properties of an unknown material, and it is desirable to know what the unknown material is based on these sensor data. In laboratory settings, this scenario is reversed and a known material is investigated for certain properties. Such an example is in the field of medical and seismic imaging where electromagnetic waves are passed through a substance and deflect in certain ways. Since certain materials are known to deflect in certain ways, these deflection data may be interpreted to discover the identity of the compositional makeup of the unknown substance. In functional terminology, let S represent a set of known substances and T a set of properties of substances in S. There is a function $f : S \mapsto T$, which means there is a function mapping a substance to its property. But in a sensor problem, such as computed tomography, we are given the property data T and must find a way to map this set back onto S. Such functions that reverse the mapping behavior of f are called inverse functions and are usually denoted $f^{-1} : T \mapsto S$. Inverse functions are not always easy to discover, particularly when the function is not injective. This is easy to see, since an image in T is not uniquely mapped. In going back to S, there may be multiple inverse images from T to S.

Proposition: Let $f_1 : S \mapsto T$ and $f_2 : T \mapsto S$. Then the following statements are equivalent:

1. f_1 has an inverse if and only if f_1 is bijective.

2. If f_1 has an inverse, then the inverse is unique.

3. If $f_1 \bullet f_2 = I_T$ and $f_2 \bullet f_1 = I_S$, then f_2 is the inverse of f_1.

4. f_1 is the inverse of f_2 if and only if f_2 is the inverse of f_1.

1.5 Graphs

It is a fact that graph theory permeates all computer science, and the best computer scientists are expert in graph theory (indeed, many of them gained importance by solving problems in algorithms of graph theory). We would be challenged to name a branch of computer science that does not use graph theory to conceptualize ideas and construct solutions to problems, and sensor integration is no exception. A distributed sensor network (DSN) is a collection of possibly disparate and intelligent sensors that are spatially distributed. Essentially, it is abstracted to a graph.

1.5.1   Definitions

A graph is a topological layout of structures called *vertices* and connections between vertices called *edges*. If an edge connects two vertices, then the vertices are said to be *adjacent* or *incident*. In some literature, vertices are also called nodes. We usually define a graph as $G = (V, E)$, where V is a set of vertices and E is a set of edges $\{(v, w) : v, w \in V$ and v and w are connected$\}$. Given two vertices, the shortest number of edges in a path between the vertices is called the *distance* between the nodes. The *diameter* of a graph is the largest number of edges that comprises a path between any two vertices of the graph. Some examples of graphs are given in Figure 1.3.

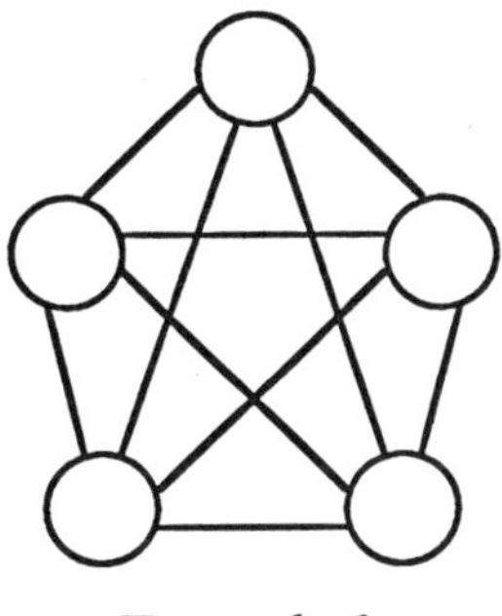

Example 1

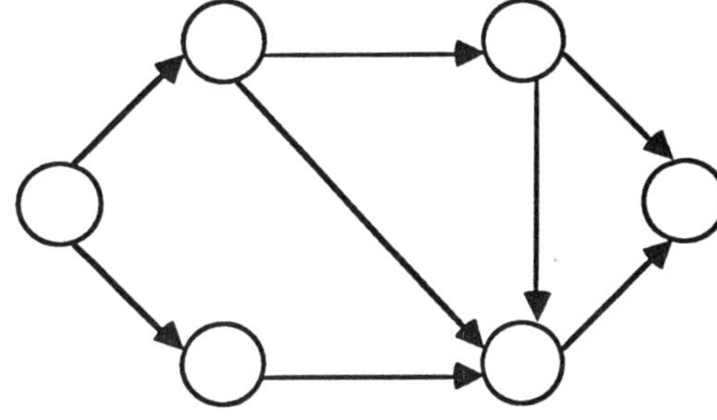

Example 2

Figure 1.3 Example graphs.

The two examples differ in a fundamental way. The graph in example 1 is said to be undirected and example 2 is said to be directed. If a graph is undirected and v and w are vertices in the graph G, then the two edges (v, w) and (w, v) are equivalent assuming that the edge exists in the graph. This equivalence is not valid in directed graphs. This concept is based on the idea of ordered sets (see above). In undirected graphs, the degree of a vertex is the number of edges that is incident with the vertex. In directed graphs, a distinction between edges that leave a vertex and edges that enter a vertex is made. The incoming degree of a vertex is the number of edges that is directed to the vertex, and the outgoing degree of a vertex is the number of edges that is directed away from the vertex.

1.5.2   Graph Isomorphisms

Not all graphs are unique. We say that two graphs $G = (V, E)$ and $G' = (V', E')$ are isomorphic if the following conditions are true:

1. Given $v \in V$ and $v' \in V'$, there is an isomorphic mapping $\pi : v \mapsto v'$ for all $v \in V$ and $v' \in V'$.

2. If $e = (v, w)$ is an edge of G, then $e' = (\pi(v), \pi(w))$ is an edge of G'.

Figure 1.4 shows two graphs of four vertices that are isomorphic.

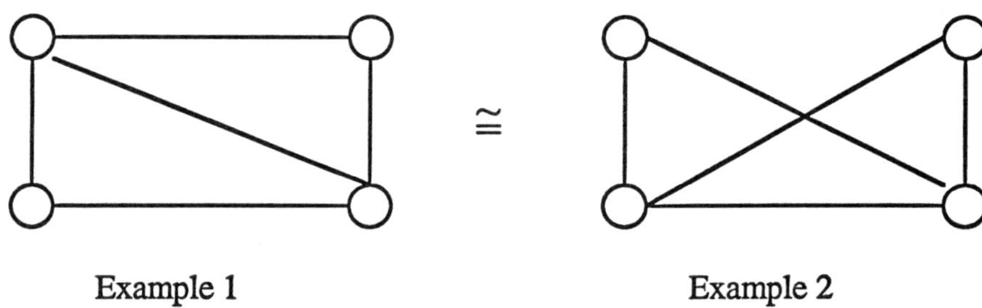

Example 1   Example 2

Figure 1.4 Isomorphic graphs.

There is an important graph called K_n, which is defined as the *complete graph* of n vertices. A complete graph of n vertices is a graph having n vertices such that each and every two distinct vertices are incident. A *subgraph* $G' = (V', E')$ of a graph $G = (V, E)$ is a graph such that $V' \subset V, E' \subset E$, and every $e' \in E'$ has end vertices in V'. Thus, every graph with n vertices is a subgraph of K_n.

Let $S = \{v_1, e_1, v_2, e_2, ..., v_n, e_n\}$ be an alternating sequence of vertices and edges of a graph, that is, $v_i \in V$ and $e_i \in E$ for $G = (V, E)$. Then this sequence is called a walk of the graph G. If the vertices in S are all distinct, the walk is called a path, and if the edges are distinct, the walk is called a trail. A graph is said to be connected if every two distinct vertices have a path connecting them. If a trail is closed, that is if $v_1 = v_n$, then the trail is called a circuit. If in the walk, $v_1 = v_n$ but the other $n - 1$ vertices are distinct, then the walk is said to be a cycle.

An important class of graphs are acyclic graphs that are directed. Such graphs have the descriptive name of DAGs (directed acyclic graphs). We comment here on them because they are useful in many areas of computer science. Some real-world applications of DAGs are in the UNIX file system, in compiler parsing strategies, and in network theory and design.

1.5.3 DSN as a Graph

A distributed sensor network (DSN) is a collection of disparate and intelligent sensors that are distributed spatially. Hence, its abstraction is a graph $G = (V, E)$, where V is a set of nodes, and E is a set of edges. For a distributed sensor network, a *node* means an intelligent node consisting of a processor and associated sensors, and an *edge* refers to the connectivity of nodes. An edge represents a communication link of the network.

For simplicity, we will treat a node as consisting of a single sensor/processor pair; although the techniques we discuss easily extend to the case for multiple sensors.

The *length* of a path between two nodes is the number of edges encountered while going from one node to another. The *distance* between two nodes is the shortest length between the nodes. The *diameter* of the network is the largest distance between any two nodes in the network. The *degree of a node* is the number of edges associated with that node. The *degree of the network* is the largest degree of any node in the network.

1.5.4 Interval Graph

We can represent a collection of N competitive abstract sensors as an interval graph. Marzullo [53] has shown that abstract sensors are isomorphic to interval graphs.

Given N abstract sensors (each expressed as an interval), we can construct an interval graph by associating each interval with a vertex and joining the two vertices if and only if the corresponding intervals intersect. For details, refer to Golumbic ([23]).

1.5.5 Information Graph

Information graphs can be used to model sensor fusion. In performing information fusion, it is necessary to identify the information available to the nodes in the network at various times and how the information at one node is related to that of another at a different time. When nodes communicate, some information is shared between them. This shared information must be carefully considered in the information integration process to avoid biased results. In particular, the shared information must be considered only once while combining information obtained from different sensors. This necessitates tracking of histories of communication between nodes and can be established conveniently using an information graph.

The information at each sensor or node in the distributed sensor network is affected by four types of events.

1. Sensor observation and transmission (IST)

2. Sensor data received at node (ISR)

3. Transmission of communication by node (ICT)

4. Reception of communication by node (ICR)

Let I be defined as $I = \text{IST} \cup \text{ISR} \cup \text{ICT} \cup \text{ICR}$. I constitutes all the significant events in the network and forms the set of information nodes in the information graph. An edge between any two nodes in this graph means that there exists a communication path between these nodes (events). By using the graph we can determine how the information flows in the system. Consider an information node $i_0 \in ICR$. This represents the event that communication from other nodes is received. Let J be the set of immediate predecessor nodes for i_0. The sensor fusion problem is then to find the information state of i_0 using the information states of nodes in J (and those of predecessor nodes of J, if necessary). Chong and associates have used the information graph to study distributed estimation problems for multitarget tracking.

1.5.6   Depth-First Search in a Graph

Given a connected graph $G = (V, E)$, consider the problem of searching G for some vertex or just visiting each vertex to perform an operation. There is a natural scheme that we may employ. Suppose we are at some vertex v. Depth-first search (DFS) proceeds by choosing a vertex w such that $(v, w \in E$, all the while keeping track of the edges incident with v that have been chosen. At w, we repeat this process with all incident vertices of w that have not been visited. Once all vertices at w have been visited, that is, there are no unvisited edges incident with w, we return to v and proceed to visit all unexplored edges incident with v. The name depth-first search is appropriate since the search of the graph proceeds by going as deep as we can from some of the initial vertex and starting over from v once we have hit "bottom." Algorithm DFS is a recursive depth-first search algorithm in the spirit of Aho et al [2].

Algorithm DFS :
Given a connected graph $G = (V, E)$ such that all vertices have been
marked as "unexplored," search the graph in a depth first search
manner from some intial vertex $v_{initial}$.
    Step 1: Mark v as "visited"
    Step 2: For all $(v, w) \in E$ do
    Step 3:   If w is "unexplored" then
    Step 4:   Mark w as "visited"
    Step 5: Repeat process for $w = v$

If the graph G is connected, this algorithm will visit all vertices in G. An alternative way to search a graph is using *breadth-first search*.

1.5.7   Trees

Another very important group of graphs is called trees. Trees are similar to DAGs in that they too are acyclic. However, trees are a restricted case, as we will see soon. The formal definition of a tree is a connected graph without cycles. Trees have an inherent hierarchical structure when the vertices contain data. In a tree there is a special vertex called the root. This is the vertex from which all paths to other vertices originate (called traversals). If the tree is directed (which most trees are assumed to be), then the root has no incoming edges. There is also a set of special vertices in a tree with no outgoing edges. These vertices are called leaves. All other vertices in a tree have exactly one incoming edge and at least one outgoing edge. If an edge $e = (v, w)$ exists in a tree $T = (V, E)$, then v is said to be the parent, or ancestor, of w and w is a child, or descendant, of v.

Spanning Trees

If $G = (V, E)$ is a graph, then there is a subgraph that is a nonunique tree $T = (V, E')$ where $E' \subseteq E$ called a spanning tree of G. There is a unique spanning tree such that if the edges of G are given values, or weights, then the total weight of the spanning tree is a minimum. Many applications, particularly network applications, must know the minimum spanning tree for a weighted graph. One algorithm is to find all spanning trees of G and pick the one that is a minimum, but this is not an efficient approach. There is an optimal algorithm that takes a greedy approach to finding the minimum spanning tree called Kruskal's algorithm. This algorithm is an inductive algorithm. Figure 1.5 shows an example of execution of the algorithm.

Kruskal's Algorithm

Given a connected graph $G = (V, E)$ and a weighting function
$f : E \mapsto R$, pick the edges as follows.
Step 1: Choose an edge e_1 such that $f(e_1)$ is minimum of all $e_i \in E$.
Step 2: If $\hat{E} = \{e_1, e_2, ..., e_i\}$ have been chosen as in step 1, choose $e_{i+1} \notin \hat{E}$ such that
    a. $\hat{E} \cup \{e_{i+1}\}$ is not a cycle of G, and
    b. $f(e_{i+1})$ is minimal of all $e_i \in E$.
Step 3: Repeat step 2 until no more edges can be added.

Binary Search Trees

There is a special set of trees that place an upper bound of 2 on the number of children a node may have. Such trees are called binary trees, and they play a very crucial role in the theory of computer algorithms. A distinction between left and right subtrees is made in binary trees. If a partial ordering is imposed on a data value of each vertex, sometimes referred to as the key value of the vertex, then the corresponding binary tree is called a binary search tree (BST). A common partial order is $\leq$. BSTs are common in algorithm design, particularly in sorting and searching problems. Figure 1.6 gives an example of a BST.

Notice that at each vertex v the left subtree has key values less than the key value at v, and the right subtree has key values greater than the key value at v. As an indication of the utility of BSTs, consider the problem of searching a list of numbers $L = \{5, 3, 4, 6, 1, 2\}$ for a particular value, say 1. To find 1 in Figure 1.6,

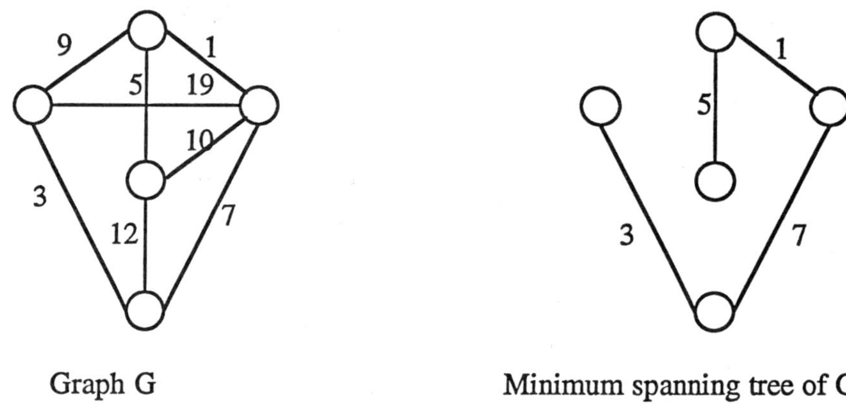

Graph G   Minimum spanning tree of G

Figure 1.5 Minimum spanning tree.

we need only search three vertices, 5, 3, and then 1. In the list L, we must travel the list to the next-to-last element to find the element 1. In general, BSTs give a search time of $O(\log n)$, whereas linear data structures require $O(n)$, where n is the number of numbers in the list L and the number of vertices in the BST for L.

1.5.8   DeBruijn Graphs

A special type of graph that is often used in distributed sensor networks in the *deBruijn graph*, denoted $DG(d, k)$. The number of vertices in a deBruijn graph is $|V| = d^k$. Also, this special graph has a diameter of k and a degree of $2d$. Many variations of deBruijn graphs have been developed for use in DSNs and multiprocessor interconnection networks, but in this book we will be interested in deBruijn graphs with a power of 2 vertices ($DG(2, k)$). Such graphs are called binary deBruijn graphs.

Each vertex has degree 4 in a $DG(2, k)$ and is given an address of k bits. Let v be a vertex of $DG(2, k)$ with address $b_k b_{k-1} \ldots b_1$. The adjacent vertices of v have addresses related to the address of v in the following way:

$$b_{k-1}b_{k-2} \ldots b_1 b_k$$
$$b_1 b_k \ldots b_3 b_2$$
$$b_{k-1}b_{k-2} \ldots b_1 b_{k'}$$
$$b_{1'} b_k \ldots b_3 b_2$$

As can be seen, the addresses are changed by shifting the end bits and complement-

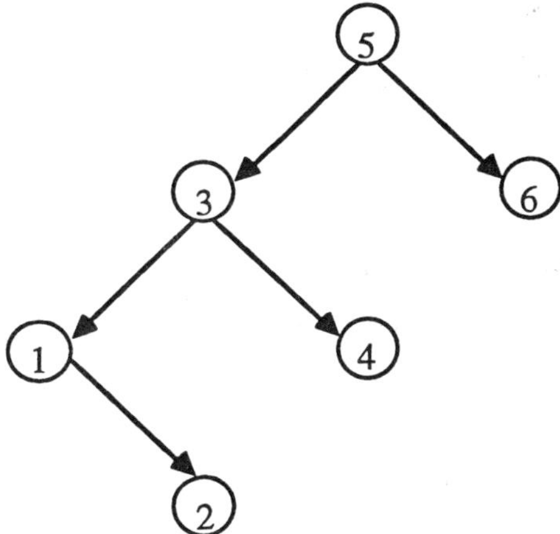

Figure 1.6 Depth-first search.

ing in two of the four cases (bit complementing is denoted by $b_{i'}$). Figure. 1.7 shows a four-node binary deBruijn graph.

1.6   Algebraic Structures

In the following discussion, some fundamental ideas concerning algebraic systems of sets are presented. The algebraic systems we are primarily concerned with are binary algebras, which are a set and a binary operation on that set. In the discussion that follows on algebraic structures, the binary operation $*$ denotes any binary operation and not necessarily multiplication. The reader is referred to Lax [46] for proof of our assertions and further reference.

1.6.1   Semigroups, Monoids, and Groups

Definition: A set S together with a binary operation $*$ on that set is called a semigroup if the binary operation is associative for all elements of the set. A semigroup is called a moniod if there exists an element e in the semigroup such that $e * s = s * e = s$ for all $s \in S$. Such an element is called an identity element. A monoid is called a group if for every element $s \in S$ there exists another element $s' in S$ such that $s*s' = s' * s = e$. s' is known as the inverse element of s, denoted s^{-1}. Frequently, a group is denoted $(S, *)$, and we will follow this convention. If $(S, *)$ is a group where the operation $*$ is commutative, then we say that $(S, *)$ is an Abelian group. Many properties of Abelian groups do not hold for general groups.

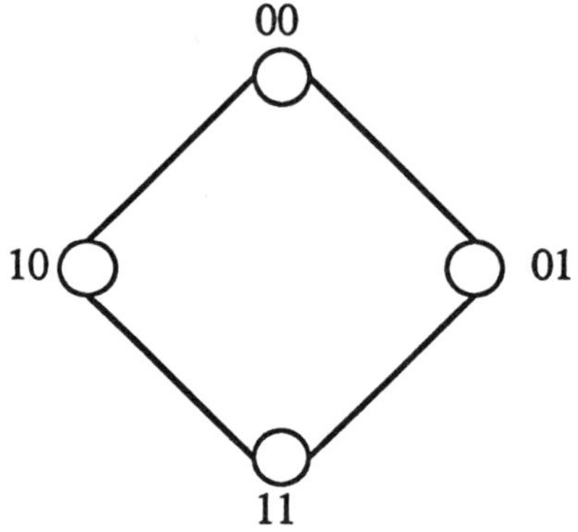

Figure 1.7 Binary deBruijn graph.

Proposition: Let $(S, *)$ be a group. Then

1. The identity of $(S, *)$ is unique.

2. The inverse of an element in S is unique.

3. $(a^{-1})^{-1} = a$ for all $a \in S$.

4. $(a * b)^{-1} = b^{-1} * a^{-1}$ for all $a, b \in S$.

5. Given $a, b \in S$, the equations $ax_1 = b$ and $x_2 a = b$ have unique solutions.

6. Given $a, b, c \in S$, if $a * b = c * b$, then $a = c$.

Examples The reader is already familiar with many of these groups. Verification of group structure is left as an exercise.

1. $(\mathbf{Z}, +)$ and $(\mathbf{R} \setminus \{0\}, *)$ are two very familiar Abelian groups.

2. $(\mathbf{Z}, *)$ is a monoid.

3. Finite-dimensional vectors of the same dimension are a group under addition.

Definition Let $(S, *)$ and $(T, *)$ be two groups. Then, a *group homomorphism* is a mapping $f : (S, *) \mapsto (T, *)$ such that if $a, b \in S$ then $f(a * b) = f(a) * f(b)$. If f is onto and one to one, then f is a *group isomorphism*.

1.6.2 Permutation Groups

Assume that X is a set and S_X is the set of all functions $f : X \mapsto X$ that are onto and one to one. This set together with the functional composition operator $f \diamond g$ forms a group known as a *permutation group* of X. If $X = \{1, 2, 3, \ldots, n\}$ where n is finite, then S_X is written as S_n. Permutation groups are not Abelian groups.

Example We consider the permutation group

$$S_3 = \{(1\ 2\ 3), (1\ 3\ 2), (2\ 1\ 3), (2\ 3\ 1), (3\ 1\ 2), (3\ 2\ 1)\}.$$

Then for functions f and g that map X to itself, we can define

$$f(1) = 2, f(2) = 1, f(3) = 3 \text{ and } g(1) = 1, g(2) = 3, g(3) = 2.$$

A more convenient notation is

$$f = \begin{pmatrix} 1 & 2 & 3 \\ 2 & 1 & 3 \end{pmatrix}, g = \begin{pmatrix} 1 & 2 & 3 \\ 1 & 3 & 2 \end{pmatrix}.$$

Functional composition is defined on permutation groups by $h(i) = f(g(i))$ or, in this example,

$$f \diamond g = \begin{pmatrix} 1 & 2 & 3 \\ 2 & 1 & 3 \end{pmatrix} \begin{pmatrix} 1 & 2 & 3 \\ 1 & 3 & 2 \end{pmatrix} = \begin{pmatrix} 1 & 2 & 3 \\ 2 & 3 & 1 \end{pmatrix} = h$$

Note that S_3 is not an Abelian group by the following example:

$$g \diamond f = \begin{pmatrix} 1 & 2 & 3 \\ 1 & 3 & 2 \end{pmatrix} \begin{pmatrix} 1 & 2 & 3 \\ 2 & 1 & 3 \end{pmatrix} = \begin{pmatrix} 1 & 2 & 3 \\ 3 & 1 & 2 \end{pmatrix} = h'$$

1.6.3 Subgroups

Definition Let $(S, *)$ be a group and $T \subset S$. Then, if

1. $a,\ b \in T \Rightarrow a * b \in T$, and

2. $a \in T \Rightarrow a^{-1} \in T$,

then $(T, *)$ is a subgroup of $(S, *)$. A more general statement would be to say that if $(S, *)$ is a semigroup, monoid, or group then $(T, *)$ is a *subsemigroup, submonoid,* or *subgroup*, respectively. That is, $T \subset S$ and T forms a semigroup, monoid, or group structure under the operation $*$, respectively.

    Proposition If $T \subset S$, then $(T, *)$ is a subgroup of $(S, *)$ iff $a * b^{-1} \in T$ for all $a,\ b \in T$.

    Proposition If $(S_1, *)$ and $(S_2, *)$ are groups and $f : (S_1, *) \mapsto (S_2, *)$ is a group homomorphism, then $(\text{Ker}(f), *)$ is a subgroup of $(S_1, *)$ and $(\text{Im}(f), *)$ is a subgroup of $(S_2, *)$.

1.6.4 Ring

Rings play an important role in many areas of applied mathematics and computer science, and the theory has its roots in cryptography and number theory. The role that ring theory plays in sensor networks and signal propagation is in error correcting codes and encoding schemes.

Definition A set R with two binary operations $+$ and $*$ is called a ring, denoted $(R, +, *)$, if $(R, +)$ is an abelian group, $(R, *)$ is a semigroup, and the elements of R obey the distributive law. If $(R, *)$ is a monoid, then $(R, +, *)$ is a ring with identity, and if $(R, *)$ is commutative, $(R, +, *)$ is an Abelian ring. If $(R, +, *)$ is a ring with identity, then the elements of $(R, *)$ that have inverses are called units. $(R, +, *)$ is called a field if the nonzero elements of $(R, *)$ form a group.

Examples Like the examples for groups, the reader already knows of many mathematical systems that are rings or fields.

1. $(\mathbf{Z}_n, +, *)$ is a ring with identity.

2. $(\mathbf{Z}_p, +, *)$ where p is a prime number, is a finite field.

3. Let $(R, +, *)$ be a ring and $M_n(R)$ be the set of square matrices of dimension n. Then $(M_n(R), +, *)$ defines a ring with identity called a matrix ring.

There are analogs of the concepts of subobjects and homomorphisms in the theory of rings.

Definition If $S \subset R$ and $(S, +, *)$ forms a ring under the same binary operations as $(R, +, *)$, then $(S, +, *)$ is a *subring* of $(R, +, *)$.

Definition If $(R_1, +, *)$ and $(R_2, +, *)$ are rings and $f : (R_1, +, *) \mapsto (R_2, +, *)$ is a mapping of the rings such that if $a, b \in R_1$ then $f(a + b) = f(a) + f(b)$ and $f(a * b) = f(a) * f(b)$, then f is a *ring homomorphism*. If f is onto and one to one, then f is a *ring isomorphism*.

1.7 Mathematical Induction

The concept of formalizing mathematical propositions is known as proof. There are many different ways of proving a statement to be true, some of which are direct proof, proof by contradiction, and mathematical induction. In discrete mathematics, mathematical induction is prevalent due to the combinatorial nature of the subject. Therefore, this section is devoted to a fundamental presentation of mathematical induction.

Formally, the *principle of mathematical induction* is defined by a subset T of the natural numbers N such that if
(a) $1 \in T$ and
(b) $n - 1 \in T$ implies $n \in T$,
then $T = \mathbf{N}$.
The use of the principle of mathematical induction in most settings implies that

T corresponds to the truth of a sequence of statements that is one-to-one with $\mathbf{N}$. This is sometimes referred to as the truth set of the statements.

To make this concept concrete, let us consider an example of how to apply the principle of mathematical induction as a proof. We will use this principle to prove the statement "If $\{x_0, x_1, ..., x_n\} \in \mathbf{P}$, then $x_0 * x_1 * ... * x_n \in \mathbf{P}$ for all $n \in \mathbf{N}$." Consider the case of $n = 1$, which corresponds to $\{x_0, x_1\}$. This is part (a) of the definition of the principle of mathematical induction and is also commonly referred to as the *base* or *degenerate* case. This case follows immediately from the order axiom. Now consider the case of $n > 1$. As an inductive assumption, let the statement be true for all $1 < j \le k$. That is, $\{x_0, x_1, ..., x_k\} \in \mathrm{P}$ implies that $x_0 * x_1 * \cdots * x_k \in \mathbf{P}$. This assumption is critical in any proof that uses induction. From this assumption, we will show that, when another element that is in $\mathbf{P}$ is added to the sequence, then the product of all numbers of the new sequence is also in $\mathbf{P}$. Indeed, consider $x_{k+1} \in \mathbf{P}$ and the sequence that results from including this new element, $\{x_0, x_1, \cdots, x_k, x_{k+1}\}$. If we multiply these elements together, we get $x_0 * x_1 * \cdots * x_k * x_{k+1}$, from which the rules of arithmetic allow us to associate in any way want. Thus, $x_0 * x_1 * \cdots * x_k * x_{k+1} = (x_0 * x_1 * \cdots * x_k) * x_{k+1}$. By the inductive hypothesis and the assumption that $x_{k+1} \in \mathbf{P}$, we conclude that $(x_0 * x_1 * \cdots * x_k) * x_{k+1} \in \mathbf{P}$ since this is simply the degenerate case. Thus, by inductive reasoning, we may conclude that if $\{x_0, x_1, \cdots, x_n\} \in \mathbf{P}$, then $x_0 * x_1 * \cdots * x_n \in \mathbf{P}$ for all $n \in N$.

On the surface, it may seem like cheating, but the inductive assumption ensures that we can add another and another number from $\mathbf{P}$ ad infinitum and still get a product that is in $\mathbf{P}$. When it is accepted that proceeding in this manner is valid and is the same as part (b) of the definition of the principle of mathematical induction, then the essence of proof by mathematical induction is discovered. The degenerate case gives a starting point for the proof.

1.8   Data Types

Commonly used data types are integers, real numbers, and complex numbers. Measurements generally have physical units (e.g., feet, degrees Celsius) associated with them. However, for some applications, relativization of data is desirable. The possibilities include the following:

Normalized data: Suppose one teacher decides to give 150 marks total for a test, while another decides to give 100 marks. How can we compare the test results? One option is to divide the marks by the total, resulting in a range from 0.0 to 1.0; such a set of values is called *normalized data.*

Fuzzy sets: In set theory, we write $a \in S$ to denote that a is an *element* or *member* of the set S. Based on the *set membership* property, set theory has operations like *set union, set intersection,* and *complement of a set.*

Fuzzy set theory allows the use of linguistic variables. Instead of treating membership as strictly 1 and 0, fuzzy set theory allows a degree of membership in a set. Thus an element x might have a degree of membership 0.6 associated with the set

A and a degree of membership 0.8 associated with the set B.

Then the *fuzzy set union* can be defined as the maximum of the two degrees of membership. x belongs to $A \cup B$ with max(0.6, 0.8) = 0.8.

The *fuzzy set intersection* can be defined as the minimum of the two degrees of membership. x belongs to $A \cap B$ with min(0.6, 0.8) = 0.6.

The *fuzzy set complement* can be defined as one minus the degree of membership. x belongs to $\bar{A}$ with $1 - 0.6 = 0.4$

Fuzzy set theory has its applications and adherents. Some results are however not intuitive, for example, $A \cup \bar{A}$ is max(0.6, 0.4) = 0.6, whereas in set theory, it is always 1 (True).

Confidence level: *Knowledge base systems* (or *expert systems*) use a set of rules for which each rule is associated with a confidence level. Then the levels of confidence might be -1 (not confident), 0 (maybe), and +1 (confident).

Probability: Probability is associated with the degree of likelihood that an event will happen due to chance. An event that can never happen has a probability of 0. A sure outcome has a probability of 1. The value of an outcome can range from 0 to 1.

Belief interval: A belief interval is used in association with plausible hypotheses and evidential reasoning (e.g. Schaffer/Dempster rule). When a priori probabilities are known, the Schaffer/Dempster rule reduces to the Bayesian method.

1.9   Computer Architecture

There are several ways to view a computer. We will describe some common techniques to classify computers.

1.9.1   Combinational and Sequential Circuits

A computer is generally made up of combinational circuits and sequential circuits. A combinational circuit is made up of logic gates such as AND (conjunction), OR (disjunction), NOT (negation) and XOR (exclusive or). The operations follow the rules of switching algebra, which is a special form of Boolean algebra. The basic logic gates can be combined into functional units, such as adders, multipliers, and comparators.

In a two-valued logic (which is most commonly used), the input and output are specified by two values: logical 0 and logical 1. The output of a combinational circuit is a function of its input values. The important point is that a combinational circuit has no memory components. There are also three-valued logic and multivalued logic.

A sequential circuit has some form of memory; hence, the output of a sequential circuit is both a function of its input values and some previous output values.

For a broader treatment, readers are referred to texts on switching circuits, digital logic, and computer organization.

1.9.2 Flynn's Taxonomy and Extensions

According to Flynn [20], computer systems can be classified by the number of instruction streams and the number of data streams available. They are called SISD, SIMD, MISD, and MIMD.

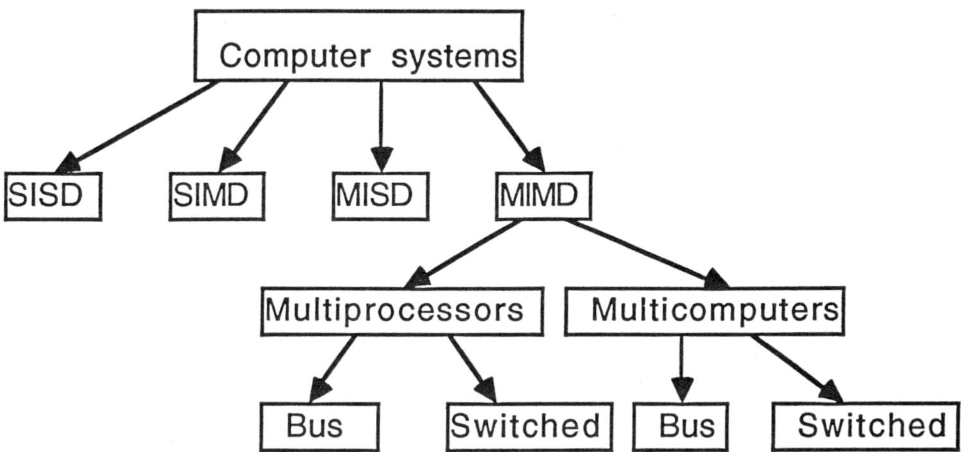

Figure 1.8 An extension of Flynn's taxonomy. Adapted from [76].

Examples of SISD (single instruction single data) machines include Intel 8080, VAX 11/780, IBM 360/91, CDC 6600, and Cray-1. Most microcomputers, mini-computers, and mainframes fall into this category. They are essentially sequential computers.

Examples of SIMD (Single instruction multiple data) machines include Illiac IV, BSP, Goodyear MPP, STARAN, and IBM 370/168. Such computers are used for vector processing. Some superminis and number crunchers fall into this category.

No known commercial computer fits into the MISD (Multiple Instruction Single Data) category.

Examples of MIMD (Multiple instruction multiple data) machines include Cm^*, IBM 370/168 (multiprocessor), Denecor HEP, and Cray-X/MP.

Tanenbaum [76] and others have extended Flynn's taxonomy(see Figure 1.8).

The class of MIMD machines can be further subdivided into multiprocessors and multicomputers. Moreover, each subclass may employ buses or switches.

The degree of coupling between processors and computers may be loose or tight. For example, multiprocessors tend to be tightly coupled, while multicomputers tend to be loosely coupled.

In shared memory systems, a portion of the memory is shared by processors. Communication can be achieved by reading into and writing from the shared memory. If several processors write into a shared memory location, there should be a convention to avoid write conflicts.

In distributed systems, we cannot assume a shared memory. A common way to communicate among processors is to send messages. Message passing is a commonly used technique. The two important questions are What to send and how to send it.

The bottom line is that there are inexpensive but powerful microprocessors at one end and number crunchers at the other end. A variety of computers (e.g., personal computers, desktop computers, workstations) can be networked to share information and data. Traditional von Neumann architecture has been supplemented by various non-von Neumann architectures (e.g. connection machine, dataflow machine, and hypercube machine).

1.10   Switching Technology

Three commonly used techniques are circuit switching, message switching, and packet switching(see Figure 1.9).

Circuit switching is analogous to making a telephone call. A path is established between the sender and receiver; it stays on for the whole length of the conversation.

Message switching is analogous to sending a parcel through the post office. The sender specifies the receiver's address and drops the parcel in a mailbox. Under normal circumstances, the parcel will reach the receiver after some time delay. Note that there may be more than one route for sending a message from the source A to destination B. If the preferred route has problems, the message may be retransmitted from A to B using another route. The key idea is that B does not have to be present when A sends the message. A common assumption is the use of a message buffer of a specified length. The message will have a header, a body, and a trailer. The header will contain the address of the sender; this is used for the reply and/or acknowledgement by the receiver.

Packet switching is a generalization of message switching. Instead of sending a very large message, it is first broken up into packets. Each packet is numbered. The packets are possibly sent along different routes and at different times. The receiver receives all the packets after some time (delay) and then uses the packet numbers to reconstruct the message. This is analogous to the case when the size of the parcel exceeds postal regulations; we break up the parcel into small packets, mark each individual packet by an identifier, and then drop the packets in the mailbox. The order in which the packets arrive at the destination is arbitrary. We assume that the receiver will ultimately receive all the packets and reconstruct the parcel using

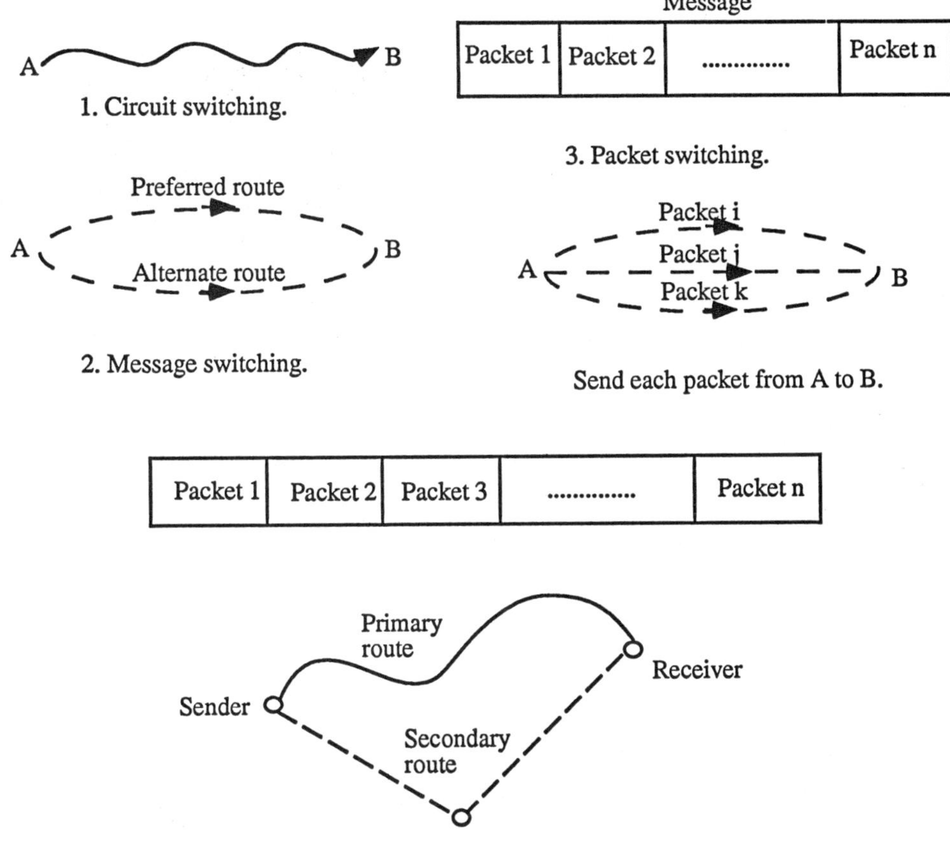

Figure 1.9 Switching techniques.

the identifiers.

1.11 Communications

Communications technology has also made tremendous advances. It has also meshed
with computer systems to a large extent. Different kinds of communication links
are used in computer systems. For connecting a terminal to a computer system,

serial communication lines supporting 1200, 2400, 4800, or 9600 bauds are used. For connecting computers over longer distances, coaxial cables, fiber optics, radio, and satellite communications are used.

Computers can be organized into networks, for example, local area networks (LAN), metropolitan area networks (MAN), and wide area networks (WAN).

In a local area network, computers within a building or several buildings within a specified radius are interconnected. A metropolitan area network covers a geographical area. At the national and international level, superinformation highways are made possible in part by wide area networks.

1.12   Intelligent Systems

We have seen the emergence of systems that have some form of intelligence, as suggested by the terms *artificial intelligence* (AI) and *computational intelligence* (CI). The term *artificial intelligence* is used in a generic sense to suggest that the intelligence is not natural but acquired. *Artificial Intelligence* (AI) comes in two types, strong and weak. Proponents of strong AI are interested in systems that mimic and/or explain human behavior. Proponents of weak AI do not require or expect a system to understand human behavior as long as it is functionally equivalent in the handling of specific tasks that require human operation or supervision.

Computational intelligence (CI), unlike AI, does not assume expert knowledge. It covers neural nets, evolutionary algorithms and fuzzy logic.

We have witnessed the widespread use of domestic appliances using some form of intelligent control, that is can deal with imprecision, uncertainty, and learning.

1.13   Distributed Systems

To paraphrase Tanenbaum [76], a distributed system is a collection of autonomous machines which are provided with a software interface that makes a user think that he or she is working on a single computer. In centralized computing, all computing is done on a single computer. Distributed systems are better than centralized systems with respect to economics, speed, inherent distribution, reliability and incremental growth. We use the term distributed as a qualifier to express some form of decentralization (e.g., distributed situation assessment, distributed diameter finding).

1.13.1   Distributed Computing

Distributed computing provides natural support for the powerful divide-and-conquer paradigm. The first step in this paradigm is to divide (or decompose) a complex task into manageable pieces called subtasks. Each subtask is then assigned to a processor (or even a computer) in the distributed system. The user need not know where the program is being run on a distributed system. The system can balance the load to improve the utilization of the system. In a distributed system, failure of a processor or a communication is not totally disastrous; the system does not crash, but degrades gracefully.

Distributed computing can reduce the bandwidth requirements. Imagine the consequences if the data communication has a limited bandwidth, and all data must be routed to a single computer over the network. In distributed computing, we are concerned with architectural issues, load balancing, scheduling, processor interconnection, deadlock prevention, and the like.

1.13.2 Distributed Problem Solving

Distributed problem solving is the cooperative solution of problems by a decentralized and loosely coupled collection of subproblem solvers (called knowledge sources) to achieve a global solution to a problem. Grofman and Owen [24] discuss information pooling and group decision making and include a review of the work of Condorcet (1785).

Based on a distributed processing domain, computations for solving a problem could be performed at various abstraction levels. Each level has a different conceptual content and generates a corresponding set of issues. Each subproblem solution at some particular abstraction level is called a knowledge source.

The knowledge sources cooperate since none has sufficient information and authority to solve the entire problem; mutual sharing of information is necessary to enable the distributed system to produce an answer as a whole.

In distributed problem solving, we are concerned with task decomposition, hypothesis testing and reduction, data fusion, and the like, on a distributed domain.

1.14 References and Further Reading

The classic text [2] by Aho, Hopcroft, and Ullman covers the design and analysis of computer algorithms. The text by Lax [46] elaborates on the algebraic structures and discrete mathematics that are covered in this chapter.

Problem Set 1

1-1. Let A={1, 3, 5, 7, 9, 11, ... } and B={0,2,4,6,8,10,...}. What is $A \cup B$, $A \cap B$, and ¬B? What might the universal set be in this example?

1-2. Let A={1,3,4,5,8} and B={2,5,9,7,3}. Give the Cartesian products, AxB and BxA.

1-3. Show that if $S \subseteq T$ and $T \subseteq S$, then S=T.

1-4. Show that a path is a trail but a trail is not necessarily a path.

1-5. Show that if T=(V,W) is a tree such that $|V| = n$, then $|E| = n - 1$.

Chapter 2

Sensor Processing

Sensor processing intuitively means processing the *raw data* provided by sensors into *information*, or useful data. Some attributes of information are as follows:

1. Timeliness

2. Accuracy

3. Precision

4. Conciseness

Sensor processing has its root in mathematics, statistics, signal processing, image processing, information theory, decision theory and control systems theory; in short, it is interdisciplinary. With advances in computer and communication technologies, sensor processing has blossomed into a discipline with important theoretical and pragmatic issues.

2.1 Motivation for Sensor Processing

Sensor processing is central to environmental systems, intelligent manufacturing systems, aerospace and defense, medical imaging, robotics, and remote systems, to name a few. The application domains are not strictly disjoint. For example, Jamshidi and Eicker [33] discuss the use of robotics and remote systems for hazardous environments. The examples given next describe the nature of sensor processing systems. They are intended to give the reader a feeling of what sensors are and how they are used in real-life applications.

2.1.1 Environmental Systems

The tremendous advances in science and technology have also seen a downside: air and water pollution, hazardous wastes, depletion of the ozone layer, disappearance of rain forests, and other ecological imbalances. An environmental system measures and monitors environmental variables such as clean air, clean water, and hazardous wastes. Figure 2.1 depicts a task structured flow of an environmental system.

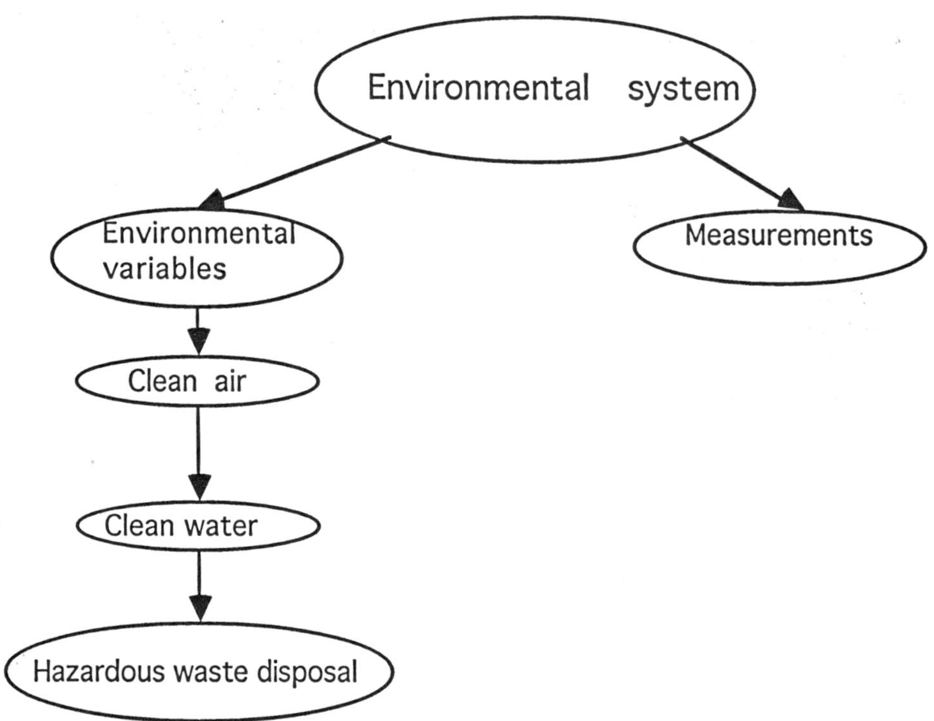

Figure 2.1 Tasks of an environmental system. Text adapted from [33].

The role of the U.S. Environmental Protection Agency is to set up and maintain programs to support the environmental laws. A primary concern is the safe disposal of hazardous wastes; using human workers requires extreme care since there can be long-term health hazards. The EPA therefore initiated a study for the use of robotics at its hazardous waste sites.

Eposito et al [19] lists some of the hazardous activities: air sampling, site assessment, soil characterization, bulk storage tanks sampling, confined space entry, soil sampling, drum sampling, groundwater sampling, and operating excavating equipment. The use of sensors is needed in most of the activities. Sensor processing is an integral component of environmental systems. This is especially true for environmental systems using robotic platforms.

2.1.2 Intelligent Manufacturing Systems

An intelligent manufacturing system is a non-deterministic control(automation) system in which, the system responds most appropriately to dynamic events taking into account the contexts, domain, and site-specific knowledge. The *site specific knowledge* can include short-term history of the soft and hard failures in the local (manufacturing) plant, ad hoc considerations that could influence resource allocation and the like. It generally means, dynamically updating the past decisions, and resorting to meritorious altenatives whic are a consequence of events that have occured recently. For example, second best routing in the case of a faulty belt on the best route.

Iyengar and Guna Seetharaman [88] have refered to a paradigm called, distributed and hierarchically organized control systems as a suitable candidate for *intelligent manufacturing systems.* For example, in a site where there are three or more robot-arms, the resource allocation, can be accomplished by a scheduling technique(deterministic) which may have to be refined in the case of a hard failure, e.g., one robot suddenly becomes less dexterous due to some reason. It can still do some job, but not the desired (optimal scheduled) job.

Manufacturing systems have evolved from fixed systems to flexible and programmable systems. The development of intelligent systems is due in part to *sensor technology* (which defines the physical limits of the design and manufacturing phases) and the integration of *soft computing, numerical optimization* and *geometric modeling.*

The term *soft computing* has been proposed by Zadeh to encompass *fuzzy logic (FL), neural nets (NN),* and *probabilistic reasoning (PR).* As he notes, FL is concerned mainly with imprecision, NN with learning, and PR with uncertainty. For problems that cannot be handled naturally by hard computing (or traditional computing), soft computing is a viable alternative.

Examples of intelligent manufacturing systems are shown in Figures 2.2 and 2.3

2.1.3 Process Control Systems

Process control systems are commonly used in industrial automation applications. A typical example is an air conditioning and heating system. A user selects the desired temperature (also called a set point) for an environment. The system uses a thermal sensor (e.g., a thermometer) to monitor the temperature of the environment. The actual temperature is compared against the desired temperature. The resulting difference (called the error signal) may or may not be negligible. If the error is small or negligible, no corrective action is taken. If the error is large, actuators are activated to affect the temperature of the environment in a way that minimizes the error. In the example, the error is provided as a feedback; the system is called a feedback control system (see Figure 2.4).

There is a variation in which several thermal sensors are used to monitor parts of the environment. In this case, the individual temperature readings are averaged and the average is compared to the set point. Averaging sensor readings is a form of sensor integration.

A control system may use fuzzy logic. A fuzzy controller conceptually consists of an analog-to-digital (A/D) converter, a fuzzifier, an inference engine, a defuzzifier,

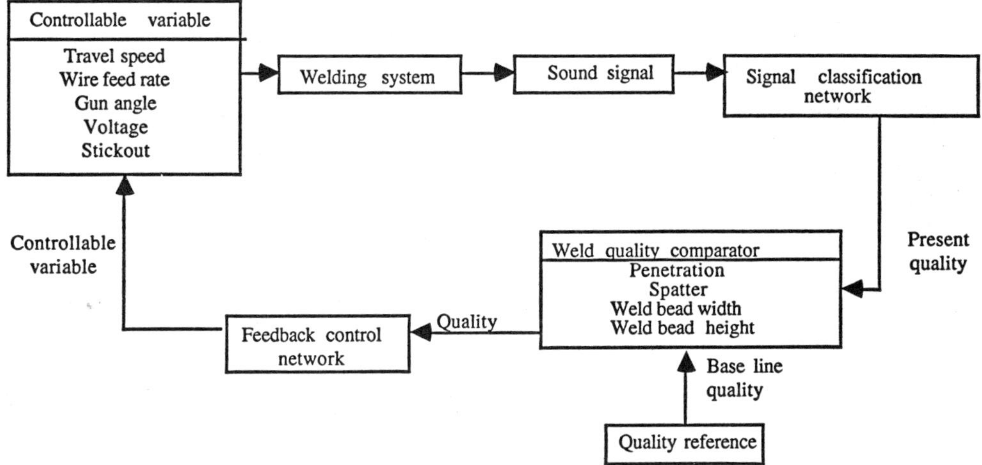

Figure 2.2 An example intelligent manufacturing system. Adapted from [89].

and a digital-to-analog (D/A) converter.

2.1.4 Surveillance Systems

Reliable detection and tracking systems are needed in peace and war. Surveillance systems form the backbone of aerospace and military applications. The ability to detect and track flying objects (e.g., planes, space shuttles, satellites, comets, UFOs, and nuclear warheads) is a huge asset. A surveillance system might minimally employ a single radar site; for reasons of security and reliability, it might be organized as several sites that can communicate with each other. Surveillance systems employ a variety of devices such as infrared sensors, microwave radars, laser radars (LADARs) and transponders.

A primary purpose of air surveillance systems is to detect and track flying objects. Transponders enable friendly planes to be distinguished from enemy planes. Radars monitor the position, velocity, and trajectory of the targets with various degrees of precision. The sensor inputs must be processed and integrated before they can be used by a military planner to devise effective countermeasures in realtime (see Fig. 2.5).

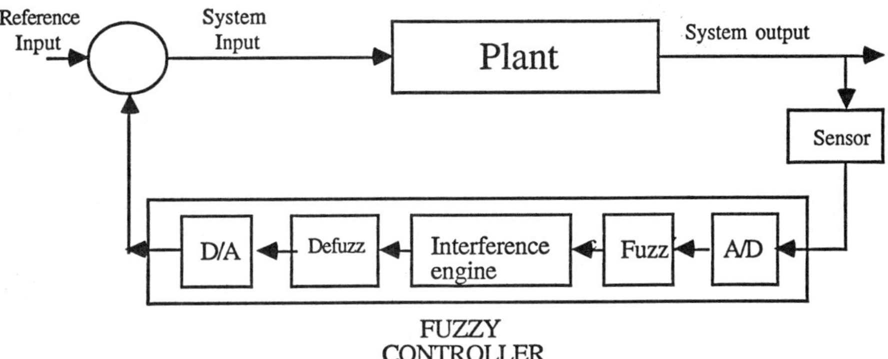

Figure 2.3 Remote systems for hazardous environments. Taken from [33].

2.1.5 Medical Imaging

There are a variety of medical imaging devices including the following:

1. X-rays (e.g. to detect broken bones)

2. Computerized tomography (e.g., to detect malignant brain tumors)

3. Ultrasound (e.g., to monitor fetal activity)

4. Nuclear magnetic resonance (e.g., to monitor blood clots)

The devices provide different kinds of data for medical diagnosis. The data
are processed by a specialist (e.g., a radiologist) and summarized into a form that
can be understood by other medical practitioners. The data taken collectively
can provide more information than the individual pieces. A set of X-rays taken
from different points of view can indicate the size of a gunwound. Since each
device presents partial information about' the human body system, these partial
information components must be integrated in a systematic and meaningful manner
into complete information to detect disorders (e.g., a broken bone or malignant
tumor).

2.1.6 Robotics

The evolution of robots has seen a wide variety of machines: stationary robots,
wheeled robots, and multilegged robots; fixed function robots and programmable
robots; tele-operated robots, tele-robots and autonomous robots. They differ in
whether locomotion function is present, whether it is designed for special or general-
purpose tasks, and the degree of human control needed to operate the robot.

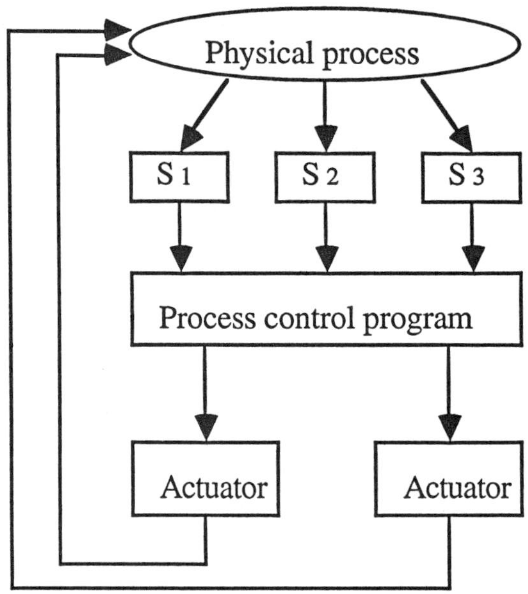

Figure 2.4 Process control sytem. The output of a physical process is monitored using n sensors. Ther process control program compares the integrated output against the desired outputand activates the actutors (if necessary) to control the physical process

All robots are equipped with a variety of sensors:

1. Contact Sensors

2. Noncontact Imaging Sensors

3. Microelectronic Sensors

4. Fiber-Optic Sensors

5. Multifunction Sensors

An *autonomous mobile robot* (AMR) is an intelligent robot capable of seeing and planning its own path in unknown and hazardous environments. As such, it requires a multitude of sensors. Unlike semiautonomous robots, which are controlled by a human operator albeit remotely, an autonomous mobile robot is endowed with

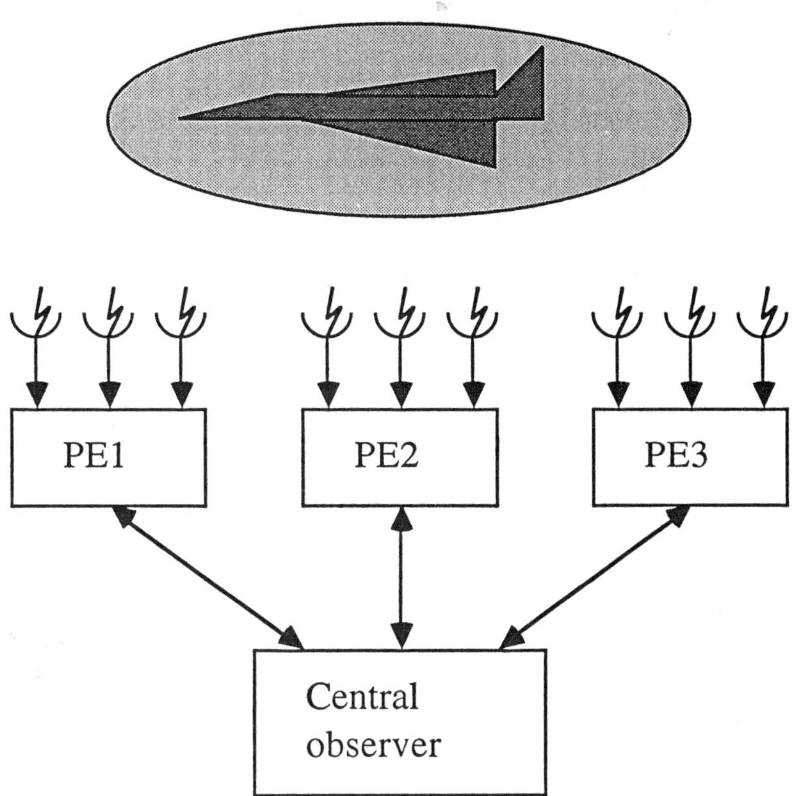

Figure 2.5 Air surveillance system

sufficient intelligence to make decisions by itself. The research on autonomous mobile robots consists of the following issues:

Rapid multimodal sensing and integration: the capability to rapidly sense external events by multiple, diverse sensors and to meaningfully integrate the information so gained

Real-time response: the capability to make decisions and take appropriate action to achieve goals without undue delay

Real-time interruptibility: the capability to interrupt normal operation and respond in a timely manner to external events occurring in its domain and then to resume the interrupted task after responding

Faulttolerance: the capability to rely on the other functional units to continue operations in the event of an internal failure

2.1.7   Remote Sensing Systems

Remote sensing systems and aerial imagery have wide applications in geological and soil mapping, land use, land cover, agriculture, oceanography, water resource planning, and other areas. In most of these applications, sensors are mounted either on satellites or in low-flying aircraft.

For example, the *Advanced Very High resolution Radiometer* (AVHRR) placed on board the NOAA-7 satellite is used to capture infrared images of the ocean. The Naval Research Laboratory Remote Sensing Division collects and processes such images for the study of ocean dynamics. The raw data present in the digital image must be processed to show the mesoscale ocean features such as cold eddies and warm eddies. They have dimensions of the order of 50 to 300 km.

2.2   Sensor

A *sensor* is a device for sensing a physical variable of a physical system or an environment. A sensor can be mechanical, electrical, electromechanical, electronic, magnetic, electromagnetic, or optical, to name a few. In fact, a sensor is any device that responds to the physical variable being monitored.

Any device that is directly altered in a predictable, measurable way by changes in a real-world parameter is a sensor for that parameter. A sensor thus has a known *transfer function* that relates the sensor's input parameter (the independent variable) to its output parameter (the dependent variable). A sensor's output parameter can thus be taken as an analog of the input parameter from the environment [29].

Any *physical system* may be viewed as a *state machine* consisting of a number of *states* and *commands* to change states; then, a sensor is a means of recording a state of that machine.

2.2.1   Physical Variables and Sensors

A *physical variable* is a variable of a physical system or an environment. Examples of physical variables are temperature, flow, position, pressure, distance, location, mass, density, radiation, light, and force. To measure them, many varieties of sensors are available. Some examples are given next.

Temperature: A thermometer is used to sense temperature. We are familiar with the clinical thermometer used to measure the body temperature. Other thermal sensors include thermocouple, thermistor, and resistive temperature device (RTD).

A thermocouple is a device that uses the *Seebeck effect* to measure temperature. When two dissimilar metals are joined to form a loop and the junctions are subjected to different temperatures, a current is produced that is proportional to the temperature difference between the two junctions.

A thermistor (or a thermal resistor) is a semiconductor device whose resistance decreases with temperature. By applying voltage across a thermistor and then measuring the current that flows through it, we can indirectly measure the temperature of the thermistor.

Resistive temperature detectors function on the principle that the resistance of certain metals increases with the temperature.

Flow: The Venturi and the turbine are used to sense flow.

Position: Microwave radars are presumably modeled after bats. Sonars are presumably modeled after porpoises. They are examples of position sensors; i.e. they are used to sense the position of an object. Infrared sensors are used in surveillance and combat planes for night time flying.

Pressure: An aneroid barometer is used to sense atmospheric pressure. Alternatives include capacitive pressure sensor and solid state gauge pressure sensor.

2.2.2   Detailed Look at a Sensor:

A barometer is a pressure sensor and can be constructed in several ways. Physics textbooks suggest the use of a calibrated mercury column to sense the atmospheric pressure acting on it. At sea level, the height of the column will be approximately 76 cm (or 30 in.). Since mercury barometers are bulky, aneroid barometers are used in airplanes and dirigibles. A key observation is that a sensor to measure a given parameter can have different forms, but all these forms must have the same functionality. Although the mercury barometer and aneroid barometer differ in their construction, their common function is to sense pressure.

How often should we sense a physical variable? Should we sense continuously, or should we sense at time intervals? If the physical variable tends to change dramatically, then the obvious answer is to sense it continuously; otherwise, it is sufficient to sense it at time intervals. The sampling rate may be uniform or nonuniform. If we do not expect abrupt changes in pressure, we can read out barometer measurements at specified time intervals.

Let $P(E, t)$ denote the pressure of the environment E at time t. The notation allows us to handle multiple sensor readings. There are two cases.

The first case involves multiple sensor readings of the same environment over a given period of time, say:

$$t_1, t_2, ..., t_n.$$

Then a sequence of barometer readings for the environment E over the period of time will be denoted by:

$$P(E, t_1), P(E, t_2), P(E, t_3), ..., P(E, t_n).$$

Collectively the readings can indicate more about weather conditions than a single isolated reading. As a rule of thumb, a gradual change of pressure indicates fair weather, while a sudden change suggests foul weather.

The second case involves multiple sensor readings of different environments, say :

$$E_1, E_2, ..., E_m, \text{ at a given instant of time } t.$$

Then the barometer readings for the different environments at time t will be denoted by:

$$P(E_1, t), P(E_2, t), P(E_3, t), ..., P(E_m, t).$$

The readings can be used to find out high-pressure and low-pressure areas. Also, the data can be used for interpolation and for drawing isobars (regions with the same pressure).

There may be interactions among the physical variables. For instance, atmospheric pressure decreases with altitude. A barometer reading inside a plane not only measures the atmospheric pressure, but can also indicate the approximate altitude. We can say that a barometer directly measures atmospheric pressure and indirectly measures altitude.

2.3   Two Aspects of Sensors

Radar is an acronym for radio detection and ranging. Radar is one of the earliest sensors used in aerospace and defense applications. Radars might employ microwaves or lasers. The term *ladar* is used for a laser radar. Bats are known to have built-in "ultrasonic" guidance systems. For night flying and combat vision, *infrared sensors* are also used. Space telescopes are indispensable tools for astronomers and astrophysicists.

Several sensors have been developed for oil and gas explorations. Seismic processing requires small explosives to be placed at strategic places; it then records the reflections of the seismic waves using *geophones* on land and *hydrophones* for off-shore and deep-sea drilling. Noise is reduced from the signals by using digital filters; the signals are then processed to extract the temporal and spatial features.

Computer systems use different kinds of sensors. For instance, most computers, even personal computers, come with a *mouse*. A mouse is a sensor for detecting relative movements of a user's hand. It is used indirectly to control a cursor on a visual display unit. An *acoustic coupler* is an example of a *transducer*, that is, a sensor for transforming one kind of signal into the other. An acoustic coupler is a way to connect a personal computer at home to a more powerful computer (say at the office) by a telephone line. A computer or a robot can be equipped with a *Charge-Coupled Device* (CCD) camera to record images.

Note that it is futile to give a comprehensive list of the sensors or the detailed workings of commonly used sensors, since they are highly dependent on technology. For our discussion, we want the reader to observe that any sensor has two interesting aspects: *physical* and *functional*.

2.3.1   The Physical Aspect

How is a sensor made? What is its form? The terms *concrete* or *physical* sensor are used to refer to a device that samples the physical state variable of interest. A barometer and a radar are examples of concrete sensors.

2.3.2 The Functional Aspect

What is a sensor supposed to do? What is its function or abstraction? The terms
abstract or *logical* sensor are used to refer to an abstraction of a reading taken by a
concrete sensor. There are several possible abstractions. We might denote a sensor
reading as a single number or as a dense interval on the real line $\mathcal{R}$. For example,
a barometer reading of an environment E at time t may be written as follows:

1. $P[E,t] = v$

2. $P[E,t] = v \pm \Delta v$

3. $P[E,t] \in v_{low} \ldots v_{high}$

The first abstraction says that v is the value of the reading. The second abstraction says that a tolerance of Δv exists. The third abstraction says that the
true value lies in an interval, whose lower and upper bounds are given by v_{low} and
v_{high}. Note that the true value need not be the midpoint of the interval. For the
case where the true value is the midpoint of the interval, the second and the third
abstractions are equivalent.

2.4 Sensor Attributes

The design of sensor processing systems depends on the choice of sensors whose
attributes match the requirements of the application. The following is a list of
some attributes of sensors.

2.4.1 Accessibility

A sensor is said to be a detector if it can determine the absence or presence of an
object or an event. Examples are metal detectors (used in airports for screening)
and smoke detectors (used to safeguard against fire hazards).

Assume that a sensor is required to sense objects in an environment. The object
may be stationary (e.g., a chair) or moving (e.g., an animal). In the context of
robots, we refer to such objects as obstacles, and the whole environment as an
obstacle space. In the context of surveillance systems, we refer to such objects as
targets.

A sensor can have partial or complete access to the environment. Partial access
occurs if the environment is too large to be handled by a single sensor. Although it
is theoretically possible to have a single air surveillance system for the 48 contiguous
states of the United States, in practice there are several surveillance systems with
overlapping jurisdictions. A sensor has partial access to the air space.

2.4.2   Dimension

A sensor is said to be linear, scalar or one-dimensional if it can be modeled by a scalar value. For example, a thermal sensor senses temperature which is a scalar value. A sensor is said to be multidimensional if it is modeled by a vector. A position sensor is a three-dimensional sensor since it senses the spatial coordinates, say (x, y, z), of an object in three-dimensional space. Likewise, a velocity sensor, say (v_x, v_y, v_z), is three-dimensional. A sensor which handles both position and velocity is six-dimensional.

2.4.3   Operating Range

Associated with each sensor is an operating range. Some sensors have a preferred operating range (e.g., short, medium or long distance). Some sensors are suitable for daytime (e.g., reliance on solar energy) or nighttime operations only; others may be operated round the clock. A clinical thermometer is designed to operate near 98.6° F, which is considered normal temperature for most people. In contrast, thermometers to monitor the internal temperature of a kiln or a nuclear reactor vessel have an operating range centered around a much higher temperature. The term *range* connotes interval width as well as the position of the interval on a value line.

2.4.4   Data

The sensor might provide continuous or discrete data. The data might be binary, numeric, or alphanumeric. The data may need to be sent to another sensor and/or a processing unit. The sensor might or might not have a buffer associated with it.

Moreover, the sensor might have to transmit the data over a communications network to a central computer for processing. The data may be organized into messages. The data may also be organized into packets. The amount of data transmitted over the network determines the bandwidth requirement.

It is common to assign a set of frequencies (or a bandwidth) for each communication channel. To reduce the amount of data needed to transmit voluminous data (e.g., a high-precision satellite image), *compression* techniques are available. Another alternative is to allow some form of *local processing* before transmission.

2.4.5   Sensitivity

A sensor is built to meet certain *specifications* such as accuracy and precision. The reading of a sensor might degrade over time or in the presence of adverse weather conditions. The degree of imprecision is also known as *tolerance*. If we say that the height of a person is 6 feet 2 inches, we normally expect a tolerance of a quarter-inch. Military specifications are often tougher than industrial specifications, because they assume the sensors are to be operated in diverse environments.

2.4.6 Location

The location of the sensor may or may not be the same as the location of processing. Such sensors are often said to be remote. Remote sensing is employed in satellite imaging, air surveillance, interplanetary-travel, instrumentation and control. The data gathered remotely must be transmitted to the processing center. The identifier of the location is also often transmitted (e.g. Geographical Information Systems).

2.4.7 Intelligence

There are several notions of intelligence. Intelligence is the ability to perform abstraction and conceptualization. In the context of autonomous intelligent systems, intelligence refers to the ability to collect information, make decisions, and implement actions that lead toward the accomplishment of some goal. [29].

A sensor is said to be intelligent if it has processing and/or decision capabilities. The use of intelligent sensors in a system allows a trade-off between computation and communication. Local processing capabilities can reduce the communication bandwidth needed for a sensor network.

2.4.8 Active versus Passive Sensors

A sensor is said to be active if it operates by emitting energy in one form or another and then detecting and interpreting any returned energy. Examples of active sensors include sonar, laser and microwave range finders [29].

A sensor is said to be passive if it simply interprets changes in some property of the environment. It does not emit energy as part of the sensing process. Examples of passive sensors include thermocouples, barometers, and photocells [29]. Note that the incidental emission of energy from a sensor (e.g., heat dissipation in a thermistor) is not taken into account in the classification of active and passive sensors.

A collection of sensors may be totally active, totally passive, or mixed (some active and some passive). If there is more than one active sensor, synchronization of the active sensors is crucial to the performance of the system.

2.4.9 Contact versus Noncontact Sensors

Contact sensing units are those which make physical contact with something in order to produce an output. Examples of contact sensors include switches, some displacement transducers, and collision detectors.

Noncontact sensing units usually operate without coming into direct physical contact with the thing or object to be sensed. Examples of noncontact sensors include photoelectric eyes, capacitive proximity sensors, and light sensors.

Certain sensors can be used as either contact or noncontact depending on the application. Thermocouples are considered contact sensors for most applications because they require direct contact with the medium whose temperature is being measured. However, thermocouples can be used as noncontact sensors for some high-temperature applications where they are exposed to the radiant energy of the sensed object [29].

2.4.10 Technology

Current sensor technologies include electro-optics, active and passive Ultraviolet to LWIR, ground penetrating radar, passive millimeter wavelength imaging, neutron activation imaging, multispectral and hyperspectral imaging and X-ray tomography.

2.5 Sensor Processing

Sensor processing is an important and central problem in signal and image processing. Sensor processing can be viewed as consisting of four activities:

1. Acquisition (data Collection, detection, or sensing)

2. Processing (local computing)

3. Integration (fusion)

4. Analysis (data interpretation or decision making)

A sensor processing system may perform some activities superficially or omit them altogether. Suppose we consider a nurse taking the body temperature of a patient. The acquisition activity corresponds to the nurse reading the thermometer. A numerical value (in either Fahrenheit or Celsius) will be returned. The analysis activity corresponds to the nurse's action. If she says that the patient's temperature falls in the normal range, she will simply jot down the value and do nothing; otherwise, she might decide to take corrective action (e.g. give medication). We might consider the example as a simple sensor processing system with only two activities: acquisition and analysis.

Most sensor processing systems will perform the four activities in the "natural" order: acquisition, processing, integration, and analysis (see Fig. 2.6). The physical variable is sensed by the acquisition activity. The raw data are then processed into a form (say scaled data) suitable for sensor integration. The integrated output is given to the analysis activity, which comes up with a decision. The decision-making process can be deterministic, stochastic, or empirical.

A sensor processing system may be organized as a hierarchy of subsystems. As a special case, consider a system organized as a complete binary tree, which consists of a root node and intermediate nodes that each have exactly two children. Assume that each node is intelligent (i.e., it consists of a processor and associated sensors).

In a centralized system, the sensor data are transmitted to a central processor to be combined. For a large amount of data, this can require huge bandwidth. In our example, since each node is intelligent, we can reduce the communications requirement by integrating the sensor data at each intermediate node and passing the result up the hierarchy. The result produced by the root node is the integrated output of all the sensor data (see Fig. 2.7).

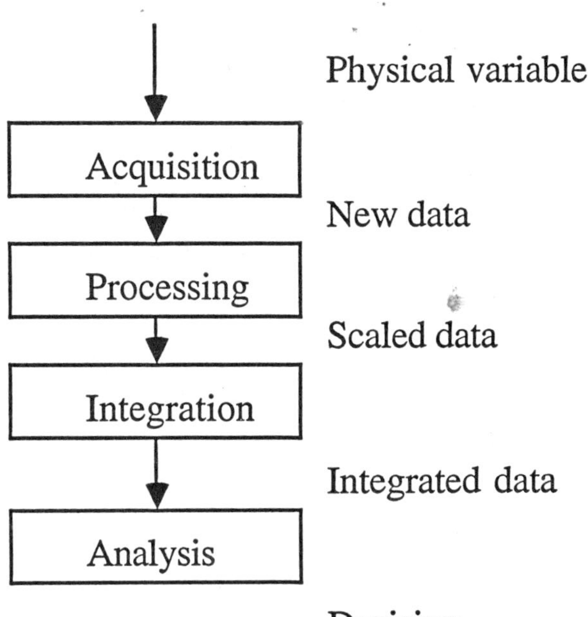

Figure 2.6 Typical sensor processing system with four activities: acquisition, processing, integration, and analysis.

2.5.1 Acquisition

The first activity in a sensor processing system is to acquire data from a physical system or an environment. We refer to the term *data* in the broadest sense. The data could be a set of measurements of a physical variable taken at a particular place over a specified interval of time. The data could also be a set of measurements of a physical variable taken at different places at a particular instant of time. In general, the data are a function of a physical variable over time and space.

One or more devices, called *sensors*, are used in acquisition. A sensor is used to sense a physical phenomenon. A *transducer* is a sensor that returns an electrical signal. A *detector* is a sensor used to detect an object or an event.

The data collected may be *continuous* or *discrete*. In *binary detection*, there are only two levels of output. It is conventional to use *one* or *zero* (corresponding to *true* and *false*). It is common practice to *sample* and *quantize* physical variables that are *analog* in nature into *digital* form; this is done in accordance with *information theory*.

The data acquired by a sensor consist of two parts. The useful part is called a *signal*. The other part is called *noise* or error. Most signals are accompanied by *white noise* which is assumed to be independent Gaussian. Various noise models

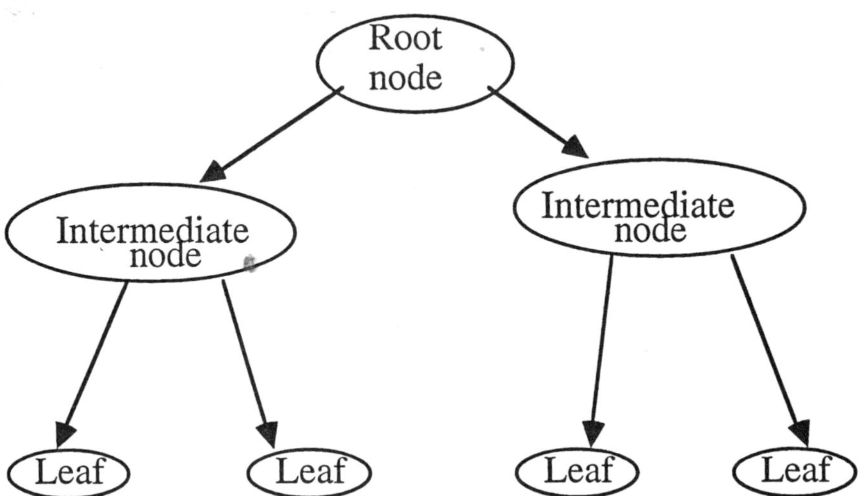

Figure 2.7 Sensor processing system in which the integration activity occurs at each intermediate node and root node.

have been proposed; they depend to an extent on the application domain.

In signal and image processing applications, the data acquired are one-dimensional and multidimensional signals. An image is an example of a two-dimensional signal.

The data collected are said to be *raw* if they are corrupted by noise and need to be refined.

2.5.2   Processing

The second activity in a sensor processing system is also known as *local computing* (see Fig. 2.8). It is also called *preprocessing* if the activity is followed by some form of *global computing*.

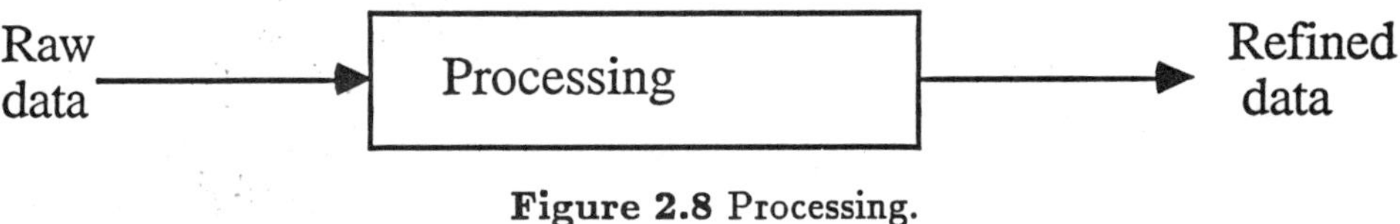

Figure 2.8 Processing.

There are several reasons why we would like to process the raw sensor data. First, the raw sensor data has a noise component; there could be some *white noise* and additional *random noise*. To improve the *signal-to-noise* ratio, an appropriate filter is used to suppress noise [1].

Second, some applications require local computing to organize the individual pieces of data (which are loosely structured or devoid of structure) into a form, which highlights a *feature*. Processing raw data can result in structures that have rich *syntactic* and *semantic* contents.

Physical units (e.g. feet or degrees Celsius) are used in measurements. For some applications, relativization of data is desirable. The use of consistent unitless representation is preferred for Integration and Analysis. Some choices are the following:

1. Normalized data

2. Fuzzy sets

3. Confidence level

4. Probability

5. Belief interval

6. An interval on a real number line

2.5.3   Integration

The third activity in a sensor processing system, integration, is also known as *fusion* (see Fig. 2.9). This activity is not needed in single-sensor system, but it is an extremely important activity in multiple sensor systems.

Figure 2.9 Integration.

The sensors can provide four types of data:

1. Competitive (redundant) data

2. Cooperative (nonoverlapping but partial) data

3. Complementary (overlapping and partial) data

---

[1] Digital filters are classified as finite impulse response filters, and infinite impulse response filters.

4. Independent (unrelated) data

Note, however, that some authors might use these terms in a slightly different manner.

Suppose we use three thermometers to measure the temperature of a nuclear reactor vessel. Then the temperature readings can be considered *competitive data*.

Suppose a region is broken up into disjoint subregions, and a sensor is used to measure a physical variable in a subregion. Then the sensor readings can be considered *cooperative data*. The data are nonoverlapping, but contribute partial information. Such subregions are also called *partitions*. For instance, we can tile(tessellate) a plane into disjoint partitions using regular polygons like equilateral triangles, squares or regular hexagons and a sensor may be placed in each subregion to obtain cooperative data.

On the other hand if we want to cover a rectangular region with circles, there is bound to be some overlap. These regions are called covers. Suppose a sensor is used to measure a physical variable for each cover; then the sensor readings can be considered *complementary data*. The data are overlapping and contribute partial information.

Two physical variables may or may not be related. They are said to be related if we can determine one from the other; otherwise they are said to be *independent*. What does it mean to integrate two independent data? There are two options. Since the data are not related, it does not make sense to combine them. Thus, the first option is to disallow or ignore such integration. If we are building and maintaining data bases, the second option calls for appending the independent data to existing records. This might require addition of new fields in a record structure.

Sensors can return contradictory and faulty data. It is imperative to combine the data in a systematic and meaningful fashion. We assume the existence of an *integrator* or *fuser* whose task is to combine the data and/or its abstractions. The following is a sample of the cases that might be handled by an *integrator*:

- No sensors fail and the distribution of the sensor readings is known

- No sensors fail, but the distribution of the sensor readings is not known

- A bounded number of senors can fail and the reliability of the sensors is known

- A bounded number of sensors can fail, but the reliability of the sensors is not known.

2.5.4 Analysis

The fourth activity in a sensor processing system is also known as *data interpretation* or *Decision making* (see Fig. 2.10). The input to a computer vision system is an image or a sequence of images. A digital image is a two-dimensional pattern of intensity values. A goal of a vision system is image understanding or scene analysis. The analysis in such a case might be to answer questions like "Is the object of type T present in the scene?" and "What objects are present in the scene?"

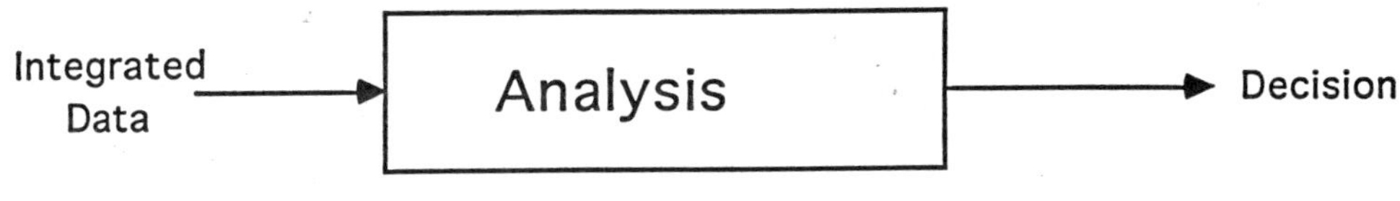

Figure 2.10 Analysis.

2.6 Distributed Sensor Processing

Sensor processing is multidisciplinary in nature. The literature is rich. Thus, we will focus our discussion on the *computational aspects* of sensor processing in general, and distributed sensor processing in particular.

Recall that sensor processing consists of four activities: acquisition, processing, integration, and analysis. One or more of the activities may take place in a distributed environment. We will call such systems as *distributed sensor processing systems*. We present a simplified example of a Distributed Sensor Processing System.

Figures 2.11 to 2.13 illustrate a distributed sensor processing system for use in an aerospace or defense application. A major objective of the system is to detect the objects (or targets) present in the observation space.

2.6.1 Task Decomposition

Assume that there are a *finite* number of resources (e.g., sensors and processors) in the distributed sensor processing system. Let us consider a system in which there are N sensors and P processors. The N sensors, say 9, can track objects in the observation space. The sensors are assumed to be the same type — microwave radar, laser radar (LADAR), or infrared — and are said to be *physically replicated sensors*. They are organized into P *clusters*, say 3, of N/P sensors each. In our example, there are three clusters, each having three sensors and one processor to control them.

The main task T is to track objects in the observation space. Consider two possibilities:

1. The observation space is too wide and thus cannot be effectively covered by any single cluster.

2. The observation space can be covered by all clusters, but a real-time response is needed to track the targets.

In the first case, we can assign the clusters to keep track of parts of the observation space. Collectively, they will cover the whole space. In the second case, we can assign each cluster to keep track of a specified number of targets; ideally, each cluster should keep track of only one target.

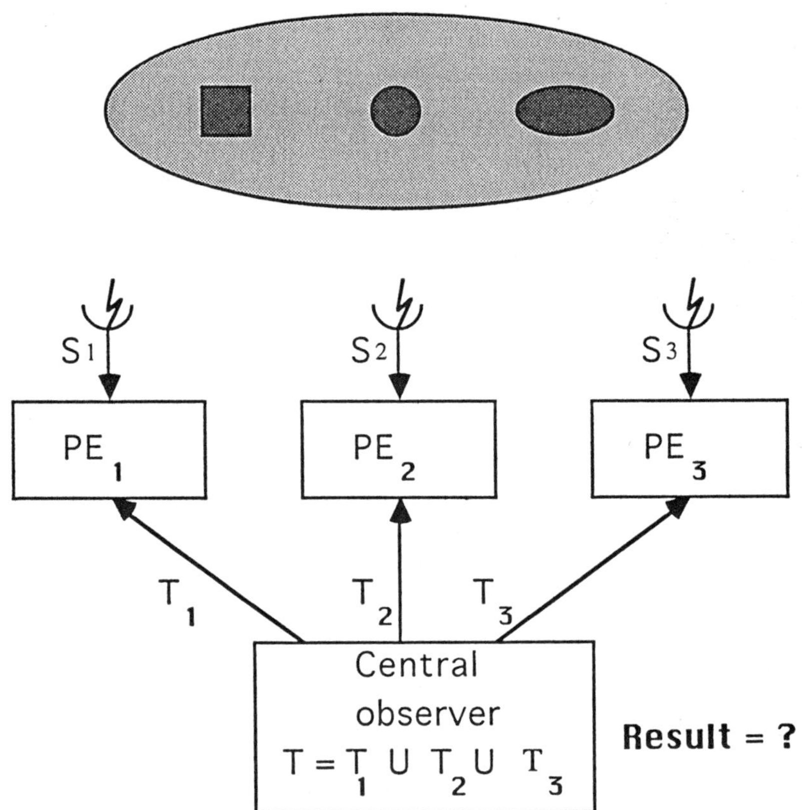

$$T = T_1 \cup T_2 \cup T_3$$

Figure 2.11 Example of task decomposition. The main task T is decomposed into three subtasks T_1, T_2 and T_3. Each subtask T_i is assigned to a processing element PE_i which controls three sensors in its cluster.

Thus, the surveillance problem is amenable to the *divide-and-conquer* paradigm. A complex problem can be conquered by first dividing (or decomposing) it into a number of subproblems that are easier to handle and then combining the partial solutions into a global solution.

In our example, the distributed sensor processing system decomposes the main task T into P subtasks; this is known as *task decomposition*.

The objective of each subtask T_i is to detect and keep track of the ith object in the observation space. Each subtask T_i is assigned to a processing element PE_i that controls three sensors in its cluster.

2.6.2 Local Computing

A cluster consists of a processor and a set of replicated sensors. The processor takes care of local computing and control; it can command the sensors in its cluster to sense the physical phenomena and return values. The data are ideally identical, but might fluctuate according to statistical distributions.

Suppose that each cluster can monitor only part of the observation space, but the targets can move anywhere within the observation space. Then the local computing phase may require communication between the sensors to share knowledge that an object has moved from one area to another.

Figure 2.12 shows local computing involving sensors $S_1 to S_3$, processors $PE_1 - PE_3$ and sensor outputs $SO_1 to SO_3$.

2.6.3 Integration

The central observer, integrator or fuser, is responsible for combining the sensor data and/or their abstractions. Note that we start off with nine sensors. There are three clusters of three sensors each. The three sensors in each cluster provide redundant data. The processor in the cluster combines the redundant data set to get a solution to a subproblem — the object being sensed by that cluster.

The observer has three sets of data, each provided by a cluster. The observer determines that there are three objects in the observation space. Note that the observer could be far away from the clusters. Then the clusters would have to transmit the result of the local computing to the observer.

In our example, we get a map of objects in the overall area of interest. Figure 2.13 the shows the integration phase involving sensors $S_1 to S_3$, processors $PE_1 to PE_3$ and sensor outputs $SO_1 to SO_3$.

2.6.4 Analysis

The system is assumed to have a knowledge base. Using the domain knowledge, the system can analyze the data and take appropriate action. In our case, the system interprets that three geometric shaped objects (corresponding to square, circle, and ellipse) are present in the observation space.

Real-life surveillance systems not only have to distinguish between friends and foes; they have to monitor the position, velocity, and trajectory of the targets and offer sufficient data to plan effective countermeasures in realtime.

2.6.5 A Complete Example

To recapitulate the workings of a distributed sensor processing system, a complete example is given. As shown in Fig. 2.11, the main task T is decomposed into three subtasks: T_1, T_2, and T_3. A subtask corresponds to detecting one object in the observation space.

Subtask T_1 is for detecting the square object. It is assigned to PE_1 (processing element 1). Subtasks T_2 and T_3 are for detecting the circular and ellipsoid objects.

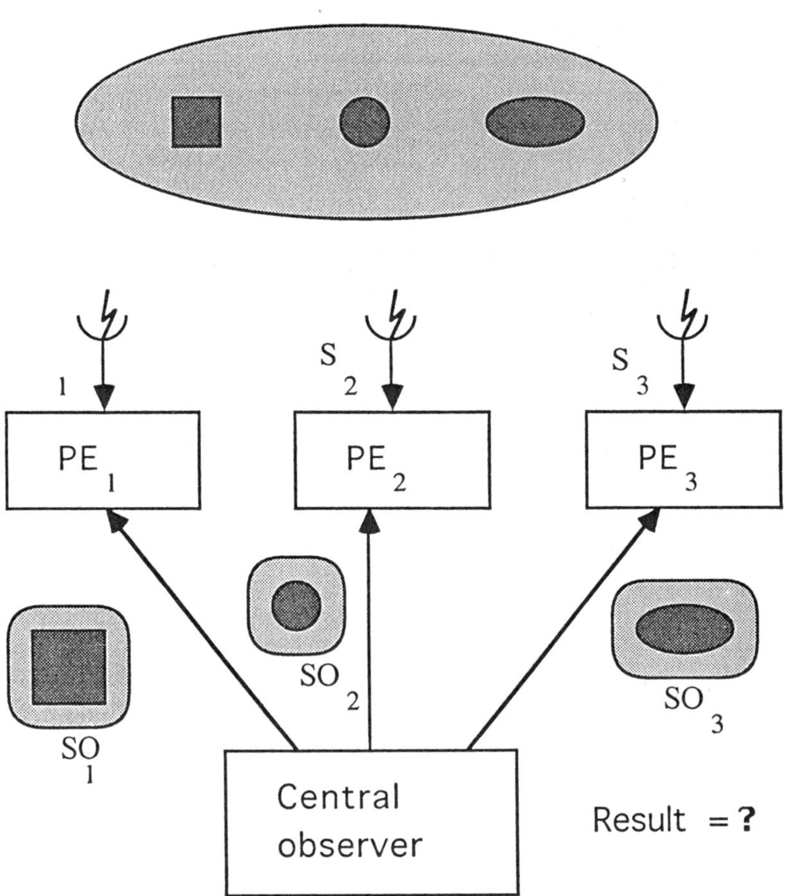

Figure 2.12 Example of local computing. $S_1 \text{to} S_3$ are the sensors. The sensor readings are processed by the processing elements to give sensor outputs $SO_1 \text{to} SO_3$.

They are assigned to PE_2 and PE_3, respectively.

Each sensor S_i, $1 \leq i \leq 3$, detects the assigned object and reports its observation SO_i, $1 \leq i \leq 3$, to the processor (Fig. 2.12). The individual sensor reports are integrated (or combined) into a single report. The resulting analysis is simply a map containing the three objects (Fig. 2.13).

2.7   Evolution of Sensor Systems

The evolution of sensor systems has seen the use of the following:

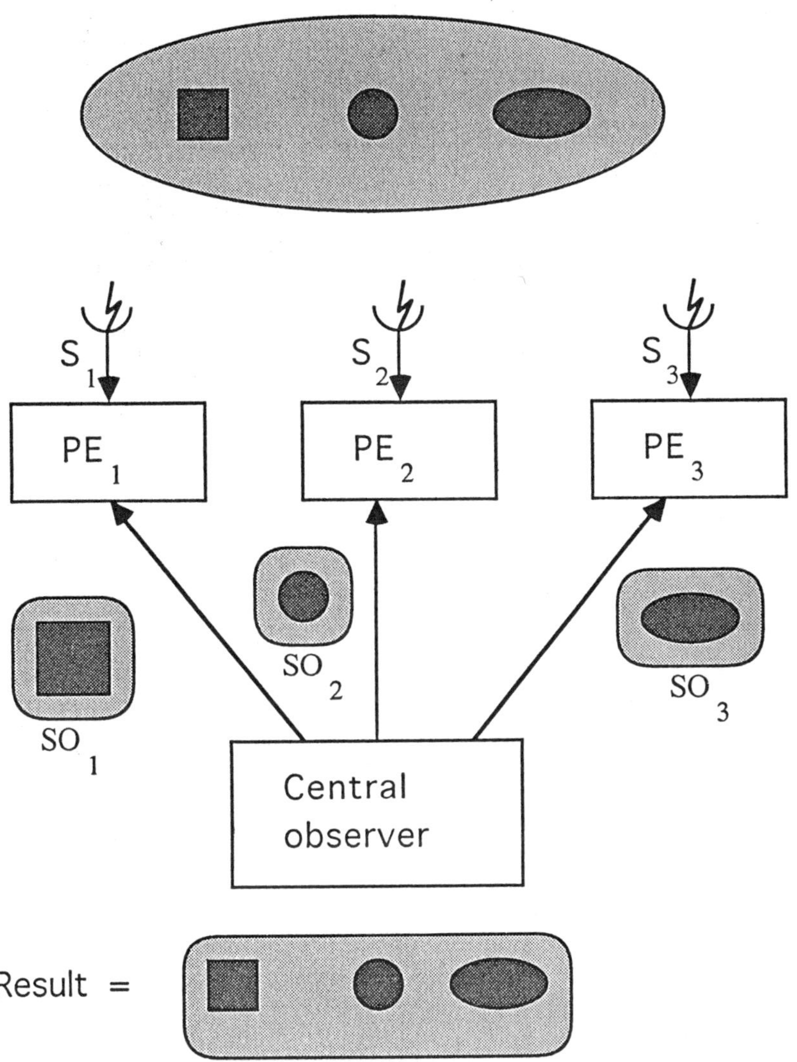

Figure 2.13 Example of integration. The central observer integrates $SO_1 to SO_3$ to get the result. The result consists of three geometric-shaped figures.

1. Single-sensor systems

2. Replicated sensors

3. Disparate sensors

4. Spatially distributed sensors

5. Intelligent sensors.

2.7.1   Single-Sensor Systems

Radar is an acronym for radio detection and ranging. A simple radar, used for target tracking, is an example of a single sensor system (see Fig. 2.13). It transmits a radio signal with a specified frequency and direction hoping to get it reflected by a target (e.g., an enemy plane). From the time taken by the signal for round-trip travel, the radar system can deduce the position of the target.

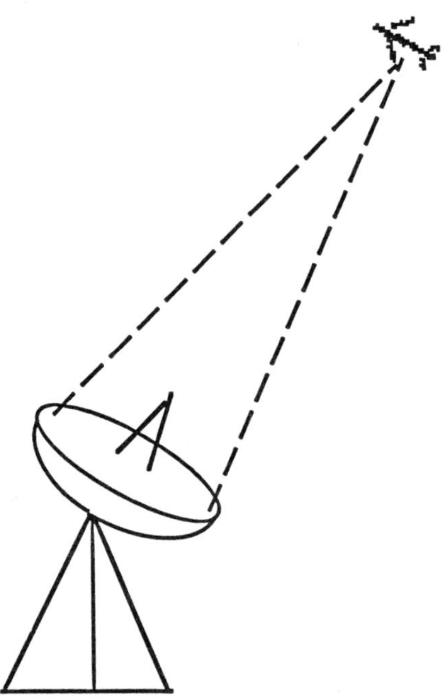

Figure 2.14 Single-sensor system. A simple radar using a single sensor, for example, microwave radar cannot tolerate failures.

Another example of a single-sensor system is sonar. It can be used to determine the depth of the ocean bed. Radars and sonars are presumably modeled after bats and porpoises which have built-in ultrasonic range finders.

Before the advent of microminiaturization, sensor systems tend to be bulky and/or expensive. Thus, such systems use single-sensors for saving space and/or money. The advantages of single-sensor systems are that they are easy to construct and analyze.

However, there are at least three disadvantages. First, single-sensor systems have limited applications and uses. For instance, an autonomous mobile robot requires several sensors (touch or tactile sensors, CCD cameras, and range finders), and thus cannot be built as a single-sensor system. Second, single sensor systems are fail hard; they are not robust for use in critical applications. Third, a single sensor cannot guarantee to deliver accurate information all the time because of its operating range and signal-to-noise ratio. Associated with each sensor is a set of limits that defines its useful operating range. For example, the electromagnetic spectrum includes gamma rays, X-rays, ultraviolet, visible light, infrared, microwave, and radio; no sensor can possibly cover the whole spectrum. Also, any sensor signal is inevitably corrupted by noise. Distributed signal processing applications require a sensor to operate in noisy environments, where the signal-to-noise ratio is small.

Thus, single sensor systems are useful for applications that are not critical, and techniques to make them robust are desirable.

2.7.2 Replicated Sensors

Associated with each device is a U-shaped curve showing the probability of the failure of the device as a function of time. A device is more likely to fail in the early stages (infancy, or burn-in period) and in the final stages (old age, or burned-out period) rather than the middle stages. Thus, in a design of single-sensor systems, it is usual to use a sensor that has passed the burn-in period. Even then, there is no guarantee that there will not be permanent or intermittent faults during operation. If time is not a critical factor, the system can and should validate the sensor readings. Based on the sensor's specification, the sensor's output can and should be ignored if it is unrealistic. An analogy is to observe the length of the shadows cast by the sun to check one's watch. For example, around noon, the shadows should disappear or be negligibly small.

Time is a critical factor in most real-life applications. Consequently, replicated sensors replaced single-sensor systems. Should there be two or three replicated sensors? If cost is a concern, two replicated are cheaper than three replicated sensors. If cost is not a concern, three replicated sensors are preferred to two replicated sensors. If there are two sensors and one of them is faulty, it is not easy to tell which one is faulty. If there are three replicated sensors, and one of them is faulty, the remaining two can guarantee a correct reading; in this sense, a system with three sensors is better.

Redundancy improves reliability by tolerating a sizable number of sensor failures. The idea of using redundant components, modules, and systems to improve reliability is common.

An example of replicated sensors is given in Fig. 2.15. The use of replicated sensors do not imply that the readings will be identical. Due to variations that are common in nature, the readings of replicated sensors will often deviate. A meaningful way to combine the replicated sensor readings is needed. Three such techniques are described next.

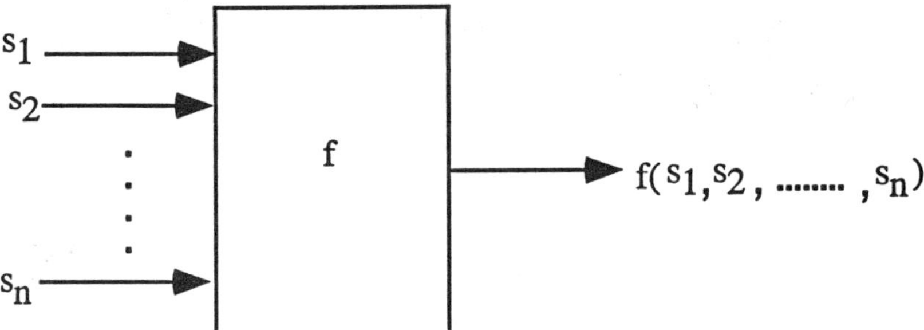

Figure 2.15 Replicated Sensors. The n replicated sensors, say $s_1, \ldots, s_n$, can improve the system fault tolerance. The sensors provide nearly identical readings. Ideally, all sensor readings should agree. The redundant readings are integrated into an output $f(s_1, \ldots, s_n)$. f is called the sensor integration algorithm.

Majority Logic

Suppose the sensor values are discrete (e.g., integer). We can compare integer values easily. The technique that we discussed earlier has a formal name, NMR (N-modular redundancy). The technique is well known and was popularized by von Neumann. N independent data are fed into a majority voter. Assuming that the good guys outnumber the bad guys, the system will return the "correct" result.

Simple Averaging

Suppose the sensor values are continuous (e.g., real). Comparing real numbers for equality is not trivial; thus, majority voting cannot be used in a straightforward manner. It is customary to use some statistical measure to interpret the N independent but not identical values. The arithmetic mean of the sensor readings can be taken to smooth out the deviation in the readings. The arithmetic mean is given by

$$f(x_1, \ldots, x_n) = \sum_{i=1}^{n} \frac{x_i}{n}$$

Weighted Averaging

This technique is also used for sensor values that are continuous. A weight (or importance) is attached to each sensor reading. The weighted readings are then averaged. The weighted mean is given by

$$f(x_1, \ldots, x_n) = \frac{\sum_{i=1}^{n} w_i x_i}{\sum_{i=1}^{n} w_i}$$

Weighted averaging reduces to simple averaging when all the weights are equal.

Example: Nuclear Reactor

The heat in a nuclear reactor is caused by nuclear fission. Cadmium rods are used to control the nuclear reaction as follows. When the reaction speeds up, the rods are further inserted into the reaction vessel to slow down the nuclear reaction. Similarly, when the reaction slows down, the rods are further removed to speed up the nuclear reaction. Thus, by manipulating the length of the cadmium rods inserted into the nuclear vessel, the reaction is controlled and a constant temperature is maintained in the vessel. It is essential to maintain a constant or bounded temperature inside the nuclear vessel, because increase in temperature beyond a critical degree results in a chain reaction that can lead to an explosion.

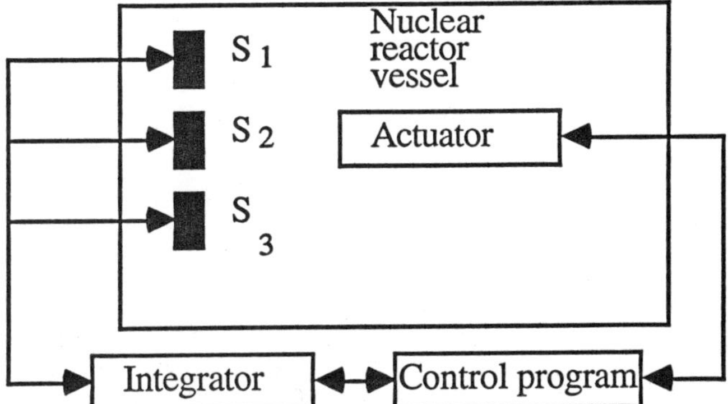

Figure 2.16 Simplified nuclear reactor vessel. The reactor uses three replicated special thermometers to improve the fault tolerance of the control system. A specified temperature range must be maintained to avert chain reaction and disaster. The integrator uses competitive integration.

Replicated special thermometers capable of measuring high temperatures accurately are used in our example. Associated with the thermometers is a common control program (CP) whose functions are as follows:

- To record the set point, that is, the temperature at which the nuclear vessel must be maintained

- To constantly sample the thermometers

- To validate the readings, that is, detect any erroneous readings from thermometers if the temperature is unrealistic

- To compute the temperature inside the nuclear vessel by combining the individual temperatures obtained from all the thermometers in some meaningful manner

- To command the actuator to position the cadmium control rods

In this example, replicated thermometers providing essentially the same data are used to increase the reliability and fault tolerance of the system. The competitive data are integrated and sent to the control system (see Fig. 2.16).

2.7.3   Disparate Sensors

A single sensor system is generally unreliable. The use of replicated sensors can improve fault tolerance. Even so, a cluster of replicated sensors has limited sensing space, and cannot often sense the complete phenomena that must be monitored. The use of one type of sensor is often not adequate for some applications (e.g. Robotics).

We will first give some examples of systems using disparate sensors (see Fig. 2.16).

Robot: An autonomous mobile robot is equipped with disparate sensors: vision, sonar, touch, infrared sensors and the like. A camera and a range finder together can provide more information to a robot in performing its tasks (e.g. path planning in unknown terrains).

Medical Imaging: X-rays, Ultrasound, Computerized Tomography (CT) scans, and Nuclear Magnetic Resonance (NMR) provide complementary methods for diagnosis.

Thematic Mapper: Satellites for remote sensing use several frequency bands. Multispectral analysis is an example of complementary integration.

Rattlesnake: It uses infrared and visual sensors to hunt for prey.

Aircraft: Reconnaisance planes and combat planes are also equipped with disparate sensors to provide complementary information.

Durrant-Whyte[17], among others, has discussed the advantages of integrating disparate sensors:

1. Increased reliability

2. Enhanced fault tolerance

3. Improved detection (and elimination) of noise.

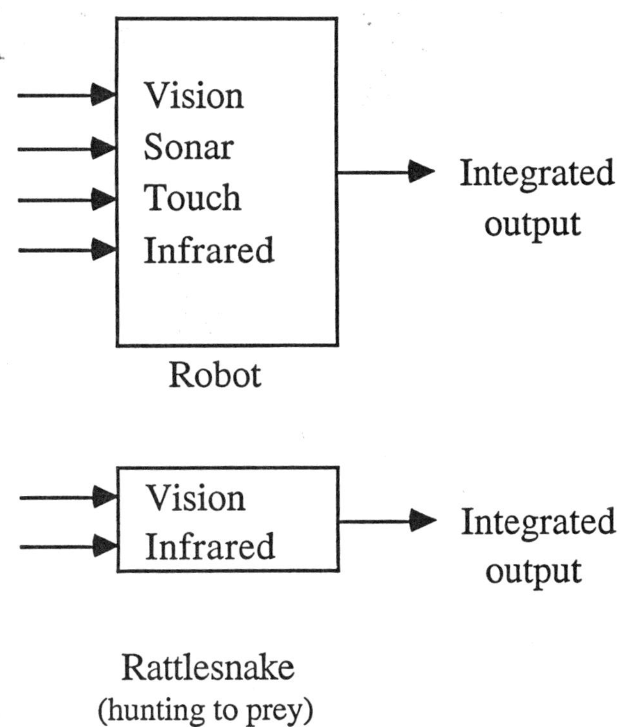

Figure 2.17 System of disparate sensors. Two or more different sensor types are involved. A rattlesnake hunting for prey uses both visual sensors and infrared sensors. A robot may be equipped with vision, sonar, touch, infrared sensors, and so on.

Noise picked up by disparate sensors tends to be uncorrelated, while the signal of interest remains correlated. Moreover, the disparate sensors can be carefully selected to complement each other; one sensor type can compensate for the shortcomings and peculiarities of another type of sensor.

2.7.4 Spatially Distributed Sensors

Some applications require that the observations of an object (or a target) be taken simultaneously from two or more points of view. In an N-nocular vision system, N cameras are placed at strategic places. The special cases where N is 2 and 3 are known as binocular and trinocular vision systems.

There are varying degrees to which sensors can be spatially distributed. The sensors may lie within a location or they may be spread over a wide region. Some target tracking systems use hundreds of sensors that are distributed geographically.

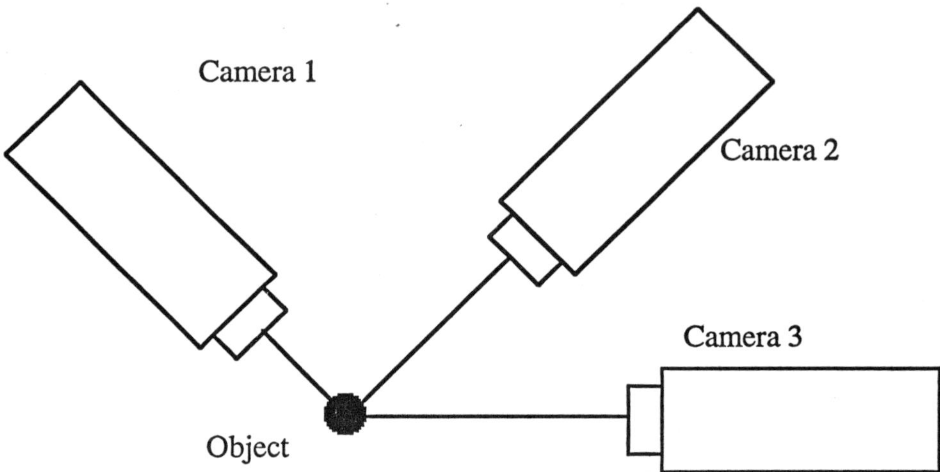

Figure 2.18 Spatially distributed sensors. A trinocular system uses three strategically placed cameras to observe an object.

Note that we do not place any restrictions on the type of sensors used. In fact, a distributed sensor can use a combination of replicated sensors and disparate sensors (see Fig. 2.18).

The data collected now varies with both time and space. *Temporal integration* and *spatial integration* are two alternatives to combine the data acquired by the spatially distributed sensors.

2.7.5   Intelligent Sensors

Recall that sensor processing consists of four activities: acquisition, processing, integration, and analysis. For a single-sensor system, processing and integration are not needed. For replicated sensors, processing might be minimal, but integration of competitive data is paramount. For disparate sensors, processing is needed to make the sensor readings compatible (e.g., the use of sensor-independent representations); integration of complementary data is needed to give a global view.

For spatially distributed sensors synchronization and communication issues must be taken into account; Integration of competitive and/or complementary data must be done. Suppose there is a large number of sensors and there is a single integration center; huge amounts of data must be transmitted from the individual sensors to the integration center. This requires a large bandwidth for data communications and a lot of work for the centralized fuser. To minimize bandwidth and centralized computing, we can use intelligent sensors (see Fig 2.19). Instead of sending huge

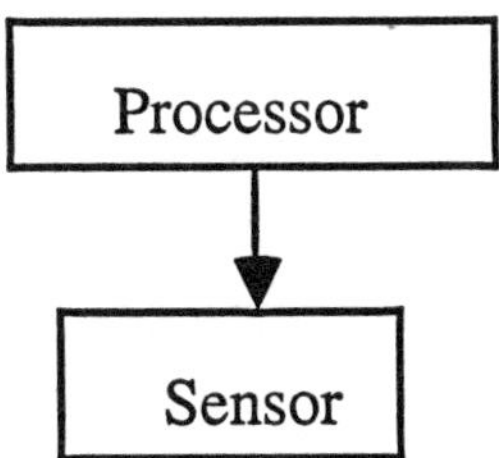

Figure 2.19 Intelligent sensors.

amounts of raw data to a centralized integration center, intelligent sensors can abstract information from the raw data before sharing the information with other sensors.

Computer technology has advanced so much that the use of inexpensive but efficient microprocessors and VLSI (Very Large Scale Integration) chips in a variety of devices (e.g., cars, television and washing machines) has become pervasive. A sensor can be made intelligent by providing processing and/or control; it is common to have a sensor/processor combination with reasonable costeffectiveness. Intelligent control systems are a fast growing discipline.

2.8 Distributed Sensor Networks

The use of intelligent, disparate sensors that are distributed spatially and geographically has become increasingly common in applications such as robotics, particle physics, medical imaging and radar tracking, to name a few. They are given a special name *distributed sensor networks* (DSNs).

A distributed sensor network is a logical step in the evolution of sensor processing systems. The design and implementation of distributed sensor networks are made possible by tremendous advances in technology, notably in computers and communications.

A distributed sensor network can be viewed as a graph $G = (V, E)$, where S, the set of vertices corresponding to the intelligent sensors (sensor/processor nodes), and E, the set of edges, describe the connectivity of the nodes.

The Cold War and the then impending danger of using nuclear warheads partially caused the U.S. Department of Defense to fund research for reliable defense systems in general and distributed sensor networks in particular. The systems were deemed feasible because of dramatic advances in computer architecture, switching technology, communications, intelligent systems, algorithms, distributed computing, and distributed problem solving. A distributed sensor Network is a set of spatially scattered, intelligent sensors designed to derive data from the environment, abstract relevant information from the data gathered, and to derive appropriate inferences from the information gained.

The need for DSNs arises in diverse applications such as intelligent robotic systems, aircraft navigation, and systems that monitor activities on an industrial assembly line, etc.

Intelligent sensors are used in distributed sensor networks. They offer several advantages.

Reduced bandwidth : A significant reduction of the bandwidth requirements of communication pathways is possible by processing signals to achieve a degree of abstraction before passing the signal on to higher levels.

Reports : Intelligent sensors can report information about their status and local conditions and perhaps even estimate the certainty (or uncertainty) associated with the signal.

Flow of information : Intelligent sensors can execute commands passed down from higher levels. This means there is bidirectional flow of information in the hierarchy.

Improved Fault Tolerance : A distributed sensor network also enjoys the advantages of the use of disparate sensors and the spatial distribution of sensors. In particular, both improve fault tolerance, since the noise tends to be uncorrelated, whereas the signals of interest remain correlated.

Efficient sensor integration : A node in a distributed sensor network consists of a *processing element* (PE) and its associated sensors. Intelligent sensors allow distributed computing at the nodes. In particular, data at a node can be integrated before passing on to other nodes. We will later describe appropriate topologies for the Distributed Sensor Networks.

Distributed Processing Techniques : A system employing multiple clusters can be organized as having:

- Centralized control and computing

- Centralized control and decentralized computing

- Decentralized control and computing

Distributed computing and distributed problem solving are commonly used in Distributed Sensor Networks.

The design of spatially distributed, target detector and tracking systems can be done by applying a distributed problem solving approach and implemented using a distributed processing system.

In a distributed environment, data and/or control can be global. We must pay attention to the information content to be communicated between various knowledge sources. We also need a knowledge source for cooperative problem solving.

In distributed processing, we are mainly concerned with the architecture issues, load balancing, scheduling, processor interconnection, deadlock prevention and the like. In distributed problem solving, we are concerned with finding problem-solving methods involving task decomposition, hypothesis, testing and reduction, data integration, and so on, on a distributed system domain.

2.9 Chapter Summary

In this chapter, we have attempted to answer the following:

- What is a sensor?

- How are sensors used in different applications?

- How can we characterize sensors?

- What is sensor processing?

- What is distributed sensor processing?

- How did sensor processing systems evolve?

- What is a distributed sensor network (DSN)?

A sensor is generally a mechanical, electrical, or electromechanical device for sensing a physical variable of the environment. Examples of sensors include thermometers, barometers, and radars. They measure temperature, pressure, and position, respectively. Sensors are used in aerospace and defense, automation, medical imaging and robotics, to mention a few areas. Sensors can be characterized by their attributes: accessibility, dimension, operating range, data, sensitivity, location, intelligence, and Mode.

Sensor processing is an important and central problem in signal and image processing. It can be viewed as consisting of four activities; acquisition, processing, integration, and analysis. One or more of the activities may take place in a distributed environment; it is then known as distributed sensor processing. An example is a distributed detection and tracking system used in aerospace and defense.

The design of sensor processing systems depends on the choice of sensors whose attributes match the requirements of the application. The evolution of sensor sysytems has seen the use of the following: single sensor systems, replicated sensors, disparate sensors, spatially distributed sensors, and intelligent sensors.

A distributed sensor network is an example of a distributed sensor processing system. It uses disparate and intelligent sensors that are distributed spatially. A DSN can be viewed as a graph $G = (V, E)$, where V is a set of intelligent nodes and E specifies their interconnection.

Even though the Cold War is over, there is still turmoil in various parts of the world and the need for efficient defense systems is far from over. In particular, a multisensor-based battle management system requires efficient sensor integration, fault tolerance, fast response, and the ability to handle countermeasures.

Moreover, the research on multisensor systems, although originated with emphasis on defense applications, is general enough for use in many other applications.

Distributed sensor networks, in general, and distributed sensor integration in particular are still active areas of research.

2.10   References and Further Reading

Some of the material from this chapter has been adapted from [9] and [28].

Problem Set 2

2-1. Differentiate the terms data and information.

2-2. Describe the following: fail hard, fail soft, fail safe.

2-3. What is an information processing system? What are the diffent types of information processing systems?

Chapter 3

An Overview of Sensor Integration

A sensor processing system consists of four activities: acquisition, processing, integration, and analysis. For a system using multiple sensors, sensor integration (or the process of combining different sensor readings) is very important.

We will first describe two examples to motivate why sensor integration is needed and what the underlying issues are. We will then lay the groundwork for subsequent chapters.

3.1 Why We Need Sensor Integration

Example 1
Prior to the use of electronic timing systems in sporting events, N (typically 3 or 5) officials were assigned to time an athlete. Due to disparities in the reflexes of the officials, the readings would seldom concur. The following scenarios show how the deviations in the individual observations might be handled (see Fig. 3.1).

1. We want to represent the output as a single number. We can use a statistical measure (e.g. mean, median, mode, minimum, maximum, or weighted mean) of the N readings. Suppose $N = 5$ and the readings are 10.2, 10.2, 10.3, 10.3 and 10.4 . Then, mean $= 10.28$, median $= 10.3$, mode $= 10.2$ or 10.3, minimum $= 10.2$, and maximum $= 10.4$, and we may choose the one that suits the application.

2. We want to represent the output as a range, that is the lowest reading to the highest reading. We write low ... high, which is read as "low to high." The difference (high to low) indicates the precision of the measurement. In our example, low $= 10.2$, high $= 10.4$, range $= 10.2 .. 10.4$, and the difference $= 10.4-10.2 = 0.2$ reflects the disparity in the speed of reflex responses of the officials.

3. For relatively large N, we can discard the lowest and the highest values and then average the remaining values. This tends to eliminate some bias in the

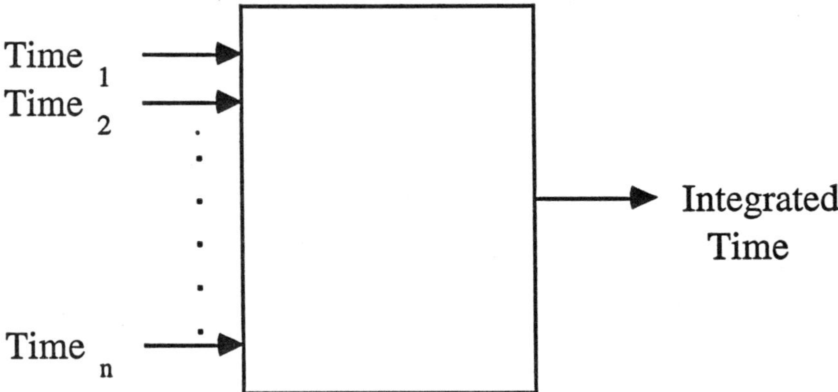

Figure 3.1 Timing in a sports event.

data set. In our example, the average ignoring the two extreme values $=$ 10.267.

4. If one reading is dramatically different from others, we can discard that reading as unreasonable and invalid. Simply average the remaining valid readings.

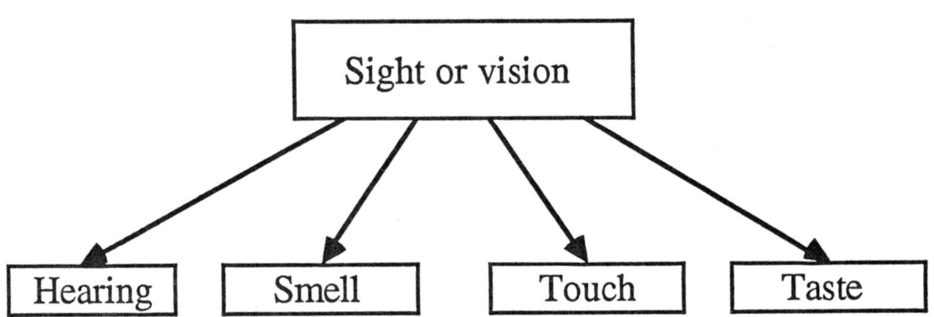

Figure 3.2 Human senses and sensors.

Example 2
Humans are endowed with five senses: vision, hearing, smell, touch, and taste (see

Fig. 3.2).

Sight or vision: Vision enables humans to interact with the dynamic environment. The two eyes, in conjunction with the brain, enable real-time image processing (stereographic vision, pattern recognition, dynamic tracking, to name a few).

Hearing: The two ears, in conjunction with the brain, can perceive directional information (speech, music and noise).

Smell: The nose can sense the various odors (fragrances and smells). The nostrils filter the air we breathe.

Touch: A function of the skin is to sense changes in temperature and humidity in the environment. Homeostasis is an example of a process control system by which the skin and the brain coordinate using feedback control to keep the body temperature "normal" (that is, around 98.6° F).

Taste: The tongue, with its taste buds, can discriminate the various flavors present in the food we eat.

The sensors corresponding to the five senses are: the eyes, ears, nose, skin and tongue. Humans also have a brain, which has been used as a model for various computing and communication paradigms. For each type of sense, the brain accepts input from the numerous sensors and nerves, filters out noise, and integrates the disparate signals in a systematic and intelligent manner to come up with the best possible interpretation or understanding.

The senses complement each other. The signals provided by two or more senses give a better overall (or global) picture. Is it a coincidence that we do not enjoy good food when we have a cold?

A decline in one sense is often compensated by another sense, as is evident in handicapped persons. The human system generally does not fail completely, that is, fail hard. It is fail soft in the sense that its performance degrades gracefully.

The input by the sensors are integrated by the brain to provide information. The raw input data are processed into useful information. The brain needs accurate, precise, timely and concise information to make important decisions as well as to handle daily chores.

The neurons cooperate and coordinate the seemingly vast and often contradictory signals, even in the presence of noise and a certain degree of failure in communications. These capabilities have been both hypothesized and studied by various researchers, notably cognitive scientists.

3.2   Sensor Integration

Sensor integration is the process of integrating, combining or fusing the sensor readings. Sensor integration techniques are important in signal processing, image processing, data communications, and consumer electronic systems. The design of efficient and robust sensor integration algorithms is crucial to industrial, medical, and aerospace and defense applications, to name a few.

We will give next some examples of sensor integration (in the broadest sense).

3.2.1 Edge Detection

In computer vision and image processing problems, the data acquired by the acquisition phase is an image, generally a two-dimensional array of intensity values(see Fig. 3.4). The image may be binary, gray level, or color. In a binary image, there are only two intensity levels: object and background. In a gray level image, the intensity of each pixel falls in the range from 0 to $2^N - 1$, where N is the number of bits (typically 8) used to represent the intensity value. There are several ways to represent color images. In the RGB representation, the red, green, and blue components of the color in each pixel are coded using N bits each (that is, a total of $3N$ bits). The raw data at the pixel level can be integrated into edges using a suitable edge detection algorithm. In a sense, edges are a better abstraction than the intensity levels of the pixels (or picture elements).

Iyengar and Deng[83] describe the importance of edge detection algorithm as follows:

"The detection of edges plays a key role in the early processing of computer vision systems, and has received considerable attention in the image processing literature. An edge in an image corresponds to an intensity discontinuity in the scene. For most machine vision tasks, an edge map is enough to carry out further processes such as motion analysis and object recognition. Edges mainly correspond to boundaries of objects in the scene. They may also correspond to the image of a shadow or a surface mark, or the effect from noise or blurring. A variety of edge detectors have been proposed (see Figure. 3.3). Most of these operators perform reasonably well on simple noise free images, but tend to fail on noisy images. In our opinion, the solution cannot be image smoothing. A better way should be the incorporation with edge contextual information.

The ultimate goal of edge detection is to characterize intensity changes in an image in terms of the physical processes that originated them[84]. It is commonly believed that, to achieve this goal, at least two stages are required, the characterization of intensity changes, and the user of structural and high-level knowledge to find the real boundaries.

Intensity changes are detected by calculating the derivatives of intensity functions, usually, the local maximum of the first order intensity differentiation or the zero-crossing of the second order intensity differentiation. The results of these differentiation operators are rough edge maps that describe intensity changes on pixels of images. Various techniques have been presented in the literature. Robert's operator[85] and Sobel's operator[86] are examples of simple edge detection operators.

Canny[87] formulated the edge detection problem as an optimization problem. He suggested good detection, good localization, and minimum flase alarms as the three objective criteria to define an optimal filter, and obtained an optimal 1-D operator for step edge detection. he found that this optimal operator can be efficiently approximated by the first derivative of Gaussian function.""

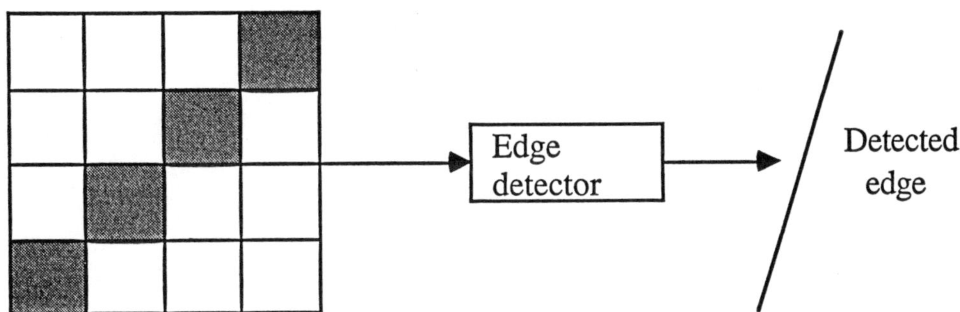

Figure 3.3 Edge detection. A binary image is represented as a grid of $N \times N$ pixels. It is input to an edge detector, which outputs the set of edges(s) it can find from the image. An edge is a kind of feature, and edge detection can be thought of as a form of sensor integration to output a feature from the given inputs.

3.2.2 Object Recognition

In recent years, there has been a tremendous spurt in the recognition of 3-D objects in range images. Visual data obtained from range sensors by a robot provides 3-D range information about objects directly. Interpretation of range data by a vision system has been one of the major problems of vision research, e.g., ability to derive properties, such as extracting features and recognizing objects. Toward this objective, we develop an efficient approach for the recognition and localization of 3-D freeform objects in range images using the properties of algebraic surfaces.

An edge can be a straight line or a curve. For simplicity, let us consider geometric objects with straight-line edges, for example, a rectangle, square, parallellogram, rhombus, or trapezium. How can we detect such geometric objects in a given image?

The first step is to extract features (e.g. edges) from the image. We also need to find the relationship between the features. For example, we need to know if two edges are equal or not equal, parallel or perpendicular and so on. The features are integrated and then matched against known objects in a data base(see Fig. 3.4). If a match is found, the algorithm reports "success"; otherwise, a "failure" is reported. Variations would involve a confidence level with which the object has been recognized.

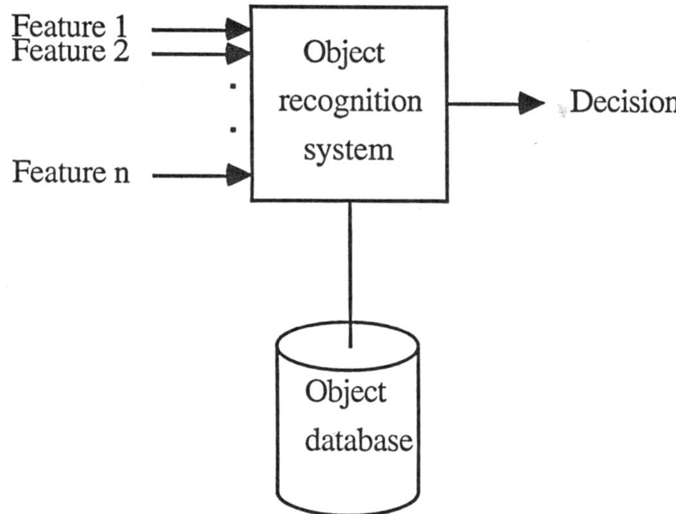

Figure 3.4 Object recognition. Given several features of a geometric object, the system matches them against an object data base to recognize the object. This involves efficient pattern matching and/or the use of inference rules.

3.2.3   Eyes of a Robot

Industrial applications use CAD (computer aided design) and CAM (computer-aided manufacturing). Industrial robots are used extensively for CAM tasks such as assembly and inspection. The earliest robots are simply of the stationary type. Later robots are mobile and can move on an industrial assembly floor by sensing obstacles and choosing a path to the destination that avoids them. Likewise, a prototype mobile robot in a hotel can move around and perform chores such as sweeping. The performance of a robot depends to a large extent on its vision — its capability to determine what-is and where-is.

Typically, there will be cameras and/or range finders that constitute the eyes of a robot. Since the obstacles are large compared with the space that can be monitored by a sensor (eye) of the robot, each sensor can only see a portion of the obstacle. The sensors are associated with a central monitor that continuously monitors the space ahead of the robot using the sensors. The sensors are regularly sampled and the partial images of the obstacles are integrated to obtain a complete picture of the space ahead (toward which the robot is moving) of the robot. If the final picture shows the presence of an obstacle in the path of the robot, the central monitor takes proper steps to direct the robot away from the obstacle.

Figure. 3.5 shows the eyes of robot. The obstacle being sensed is of rectangular shape. Each of the four sensors in the robot observes a portion of the obstacle. The

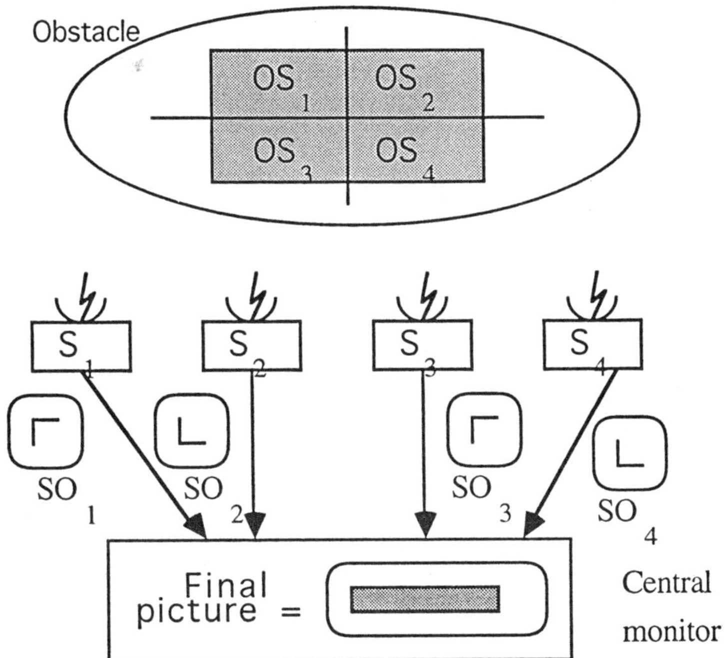

Figure 3.5 Eyes of a robot. A simple robot vision system uses four sensors to monitor the four quadrants of a workspace. This is a form of distributed detection. OS_1 to OS_4 are portions of the obstacle space. S_1 to S_4 are the sensors. SO_1 to SO_4 are the sensor outputs.

central observer (or the fuser) combines the partial images to generate the complete picture (in this case the rectangle). It is interesting to note that the final picture will be incorrect if any of the sensors fail.

3.2.4 Aerospace and Defense Sensing

Ever since radar was developed for detecting enemy planes, advances in technology have seen the emergence of sophisticated detection and tracking systems that uses a large number of different types of sensors, (e.g., infrared sensors, microwave radars and laser radars). Moreover, the sensors have different specifications, tolerances, and lifetimes and they may be distributed geographically (even worldwide). For real-time systems, reliability, precision, and efficiency are paramount. They cannot fail hard. They may fail soft (or degrade gracefully) under certain circumstances. For increased fault tolerance, replicated sensors are used in each cluster, which is responsible for handling a specific task.

In chapter 2, we described a simplified air surveillance system using nine sensors S_1 to S_9 and three processors PE_1 to PE_3. Each cluster consists of a processor controlling three sensors. Each cluster is assigned to track only one specific target (among the multiple targets), and any information it may collect about other targets is not utilized. To obtain a global (or overall) picture, the complementary information from the sensor clusters is integrated in a systematic and intelligent manner.

Sensor integration is the process of combining the information from the sensors. When the sensors are distributed spatially, the process of combining the information from the sensors is called distributed sensor integration. The air surveillance example shown in chapter 2 requires efficient distributed sensor integration to keep track of the moving targets in real time.

3.2.5　Sensor Integration in a DSN

A distributed sensor network uses disparate and intelligent sensors that are distributed spatially. Each sensor returns a value of the physical variable in the environment. The signals obtained directly from the sensors cannot be used for the following reasons.

1. Each sensor has limited accuracy. There is an uncertainty associated with the signal, and it is increased by communication delays. Different sensor signals will have different uncertainties associated with them.

2. The fusion center may be interested in a value at the time the sensor was not sampled. The value may be interpolated, but this requires complete knowledge of the physical process.

3. The sensor might have some useful properties built into it that can be extracted by processing the sensor output.

For these reasons, we need to obtain an abstraction of the physical measurement. We will call it an abstract sensor estimate, or simply an *abstract sensor*. We want to distinguish it from the physical device that made the measurement; we will call the device a *concrete sensor*.

Assume that we have abstract sensors and want to integrate (or combine) them. We have to resolve several issues.

1. Different sensors provide different sorts of information. We should decide how to represent and interrelate such varying information. Appropriate representation is a key to problem solving in general, and sensor integration in particular.

2. Different sensors take their measurements from different points. We should decide how to register the measurements; this can be done with the use of

transformations. The problem is to overcome the uncertainties in the transformations between sensing frames.

3. Sensor measurements are affected by noise. In particular, no single sensor measurement can be wholly relied on. It is possible to reduce errors using multiple measurements; however, multiple measurements can introduce contradictions.

4. Sensors and sensor interpretation programs have time constants that vary considerably. Our problem is to decide when and where to deploy each sensor.

5. Sensors fail or violate their operating limits. The set of operating sensors can also change dynamically. We need to design a robust technique for integrating such sensors.

A distributed sensor network requires some form of combining data in a meaningful manner. Ways of combining competitive, complementary, cooperative, and independent data will be covered later.

A DSN allows a hierarchy of sensor integration to be implemented efficiently. There can be at least two levels of sensor integration: intra-integration and inter-integration.

Sensor integration within a node: The sensors are generally replicated physically or logically. Each sensor ideally provides identical information. Replication offers fault-tolerance as in N-module redundancy technique. In practice, however, due to noise and other factors (e.g., communication delays), the information provided by each sensor may vary. It is possible that some sensors might fail. Fault-tolerant integration of replicated sensors is important.

Sensor integration between nodes: If two nodes use different types of sensors (e.g., radar and infrared sensor), the data from each node must be transformed into a common medium before they can be integrated. Integration of sensors from multiple nodes is also important.

3.3 Types of Sensor Integration

Sensors can provide competitive, complementary, cooperative, or independent information. To best integrate the individual pieces of data into a meaningful decision (or action), we would need one or more of the following types of sensor integration.

3.3.1 Competitive Integration

Redundant (or competitive) information allows us to construct a fault tolerant and more reliable result from the sensor readings. Competitive integration requires replicated sensor readings, which ideally are identical, but in reality may be noisy and subject to statistical variations. Even if some sensors are faulty, the information from the erroneous sensors can be ignored during the integration process.

3.3.2 Complementary Integration

Suppose partial and overlapping (or complementary) information is available from each sensor. The partial information must be integrated into complete information. If some sensors are faulty, the result of the integration can be erroneous.

Complementary integration means combining partial information to get a global picture. The information may be provided by sensors with different orientations (or viewpoints). The information may come from disparate sensors.

3.3.3 Cooperative Integration

Suppose partial and nonoverlapping (or cooperative) information is available from each sensor. The partial information must be integrated into complete information. No sensor can be faulty; the result of the integration would be erroneous.

3.3.4 Independent Integration

Suppose unrelated (or independent) information is available from the sensors. Then integrating them simply means adding more fields to a data record.

3.3.5 Temporal Integration

This requires integration of information over a given time. The data, sampled at time intervals, must be processed to re-create the notion of time sequence.

3.3.6 Spatial Integration

This requires integration of information over a given space. For example, to perceive an object in a scene, we get a number of sensor images from many vantage points. Any conclusion reached from this is what is called spatial cues or spatial inference.

3.4   Organization of Sensor Processing Systems

Sensor Processing consists of four activities; acquisition, processing, integration, and analysis. There are several ways to organize a sensor processing system. Some variants are shown in the following subsections.

3.4.1 Levels of Integration

If a sensor processing system needs to acquire a huge amount of raw data and process them into useful information, it is convenient to use a hierarchy of transformations. It is common to have three levels: lowlevel, intermediatelevel, and highlevel. For computer vision problems, the three levels correspond roughly to the following:
Low level: Conversion of intensity (pixel) map into features such as lines and edges. **Intermediate level:** Pattern recognition **High Level:**   object identification

A commonly used classification scheme is as follows:

- Data integration

- Feature integration

- Decision integration

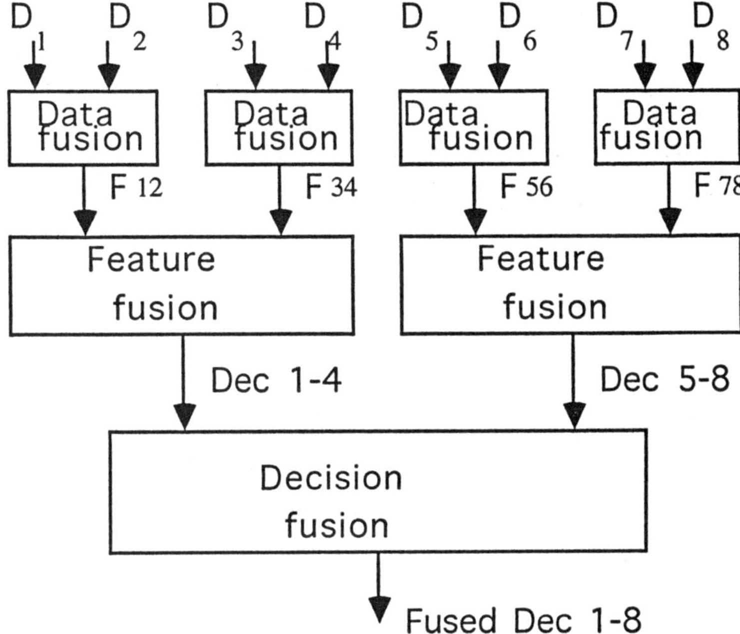

Figure 3.6 Hierarchy of sensor integration (adapted from Dasarthy). There are three levels of sensor integration: low-level, intermediate level and high-level. They are also known as data integration, feature integration and decision integration.

Consider the example shown in Fig. 3.6. The data D_1 and D_2 are combined in the data integration step into a feature F_{12}. Likewise, D_3 and D_4, D_5 and D_6, and D_7 and D_8 are combined into features F_{34}, F_{56} and F_{78} respectively. The features F_{12} and F_{34} are combined in the feature integration step into a local decision Dec_{1-4}. Likewise, F_{56} and F_{78} are combined into decision Dec_{5-8}. The local decisions Dec_{1-4} and Dec_{5-8} are combined in the decision integration step into a global decision $FusedDec_{1-8}$.

In the previous discussion, we assume that the inputs are of the same type. The type of integration is determined by the input type. Some authors consider the terms sensor integration and sensor fusion to be synonymous.

3.4.2 Integration Specified by Input/Output

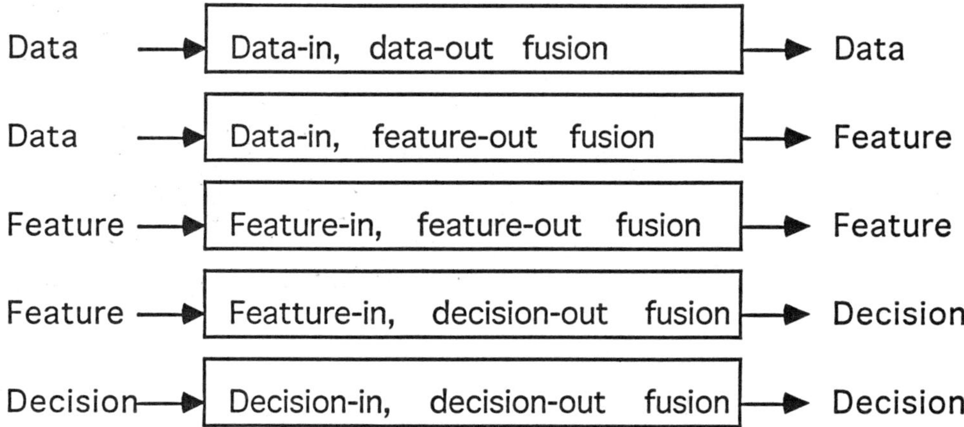

Figure 3.7 An alternative classification scheme of sensor integration (adapted from Dasarthy). It is based on the type of input and output involved in the integration process.

If we specify the integration process explicitly in terms of the ordered pair $<Input, Output>$, then five such pairs are possible(see Fig. 3.7:

- $< Data, Data >$

- $< Data, Feature >$

- $< Feature, Feature >$

- $< Feature, Decision >$

- $< Decision, Decision >$

For example, $< Data, Feature >$ means that the inputs are all of type *Data*, while the output is of type *Feature*.

3.4.3 Different Input Types

We may also consider whether the input types can be different. For instance,

- Data-data integration

- Data-feature integration

- Data-decision integration

Here, the term $Data - Decision$ means that one input is of type $Data$ and the other input is of type $Decision$.

3.4.4　Tree-structured Sensor Processing System

Another variant of a sensor processing system is shown in Fig. 3.8, where the sensors s_1 to s_7 (of the same type and specifications) are arranged as a complete binary tree. There are four *leaf* nodes, two *intermediate* nodes, and a *root* node. The intermediate and root nodes are intelligent in the sense that they can perform local processing on three input values (from two children and itself) and pass the output value to its parent. Thus, the output of the root node depends on the function F used for local processing.

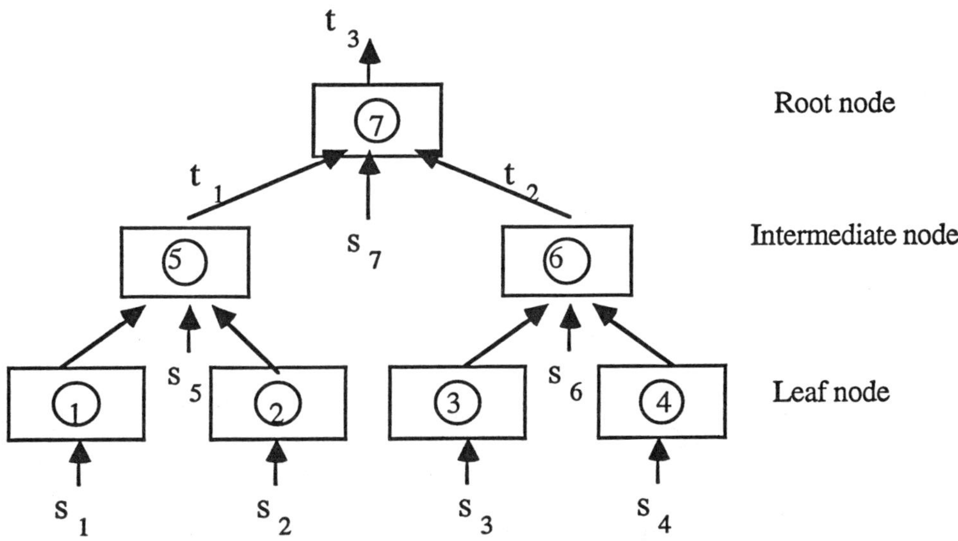

Figure 3.8 Sensors arranged as a complete binary tree. There are three levels: s_7 is the root node, s_5 and s_6 are the intermediate nodes, and $s_1 to s_4$ are the leaf nodes. Local processing involves three sensors: a parent and two children.

Processing the seven sensor readings require three sequential steps:

1. $t_1 = F(s_1, s_2, s_5)$

2. $t_2 = F(s_3, s_4, s_6)$

3. $t_3 = F(t_1, t_2, s_7)$

The final output is $t_3 = F(F(s_1, s_2, s_3), F(s_3, s_3, s_6), s_7)$. Note that the steps 1 and 2 can be done in parallel. In general, processing of n sensors arranged as a binary tree requires $\Theta(\log n)$ time.

3.4.5 Multisensor Integration

Figure. 3.9 shows S_1, S_2, S_3 and S_4 (not all of the same type). Assume that S_1 and S_2 are the same type, S_3 of a second type and S_4 of a third type. This sensor processing problem requires some form of preprocessing to make sure that the sensor data are compatible.

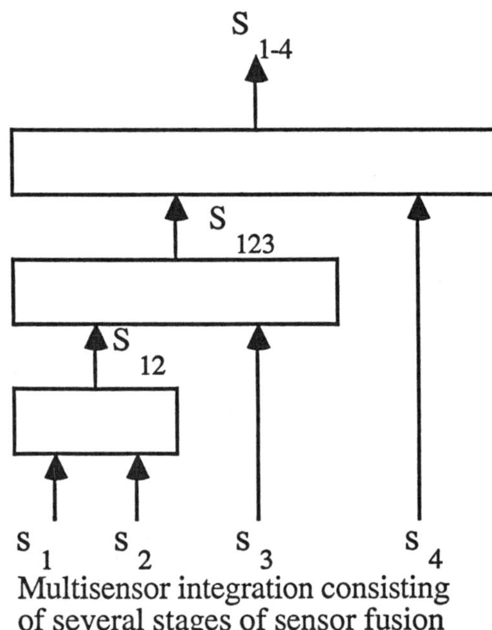

Figure 3.9 Multisensor integration (adapted from Luo and Kay). There are three types of sensors: s_1 and s_2 are of one type s_3 of a second type and s_4 of a third type. Processing them requires three sequential steps.

Processing the four sensor readings require three sequential steps:

1. $t_1 = F_1(s_1, s_2)$

2. $t_2 = F_2(t_1, s_3)$

   3. $t_3 = F_3(t_2, s_4)$

The final output is $t_3 = F_3(F_2(F_1(s_1, s_2), s_3), s_4)$. Note that the function F_1 uses competitive data, while functions F_2 and F_3 use complementary data.

3.4.6 Suite of Sensors

A collection of sensors placed in a particular configuration is called a *suite*. Sensors can be placed in series. The serial configuration is also known as a *tandem*. Sensors can be placed in parallel. Furthermore, there can be serial/parallel and parallel/serial configurations, as shown in Fig. 3.10.

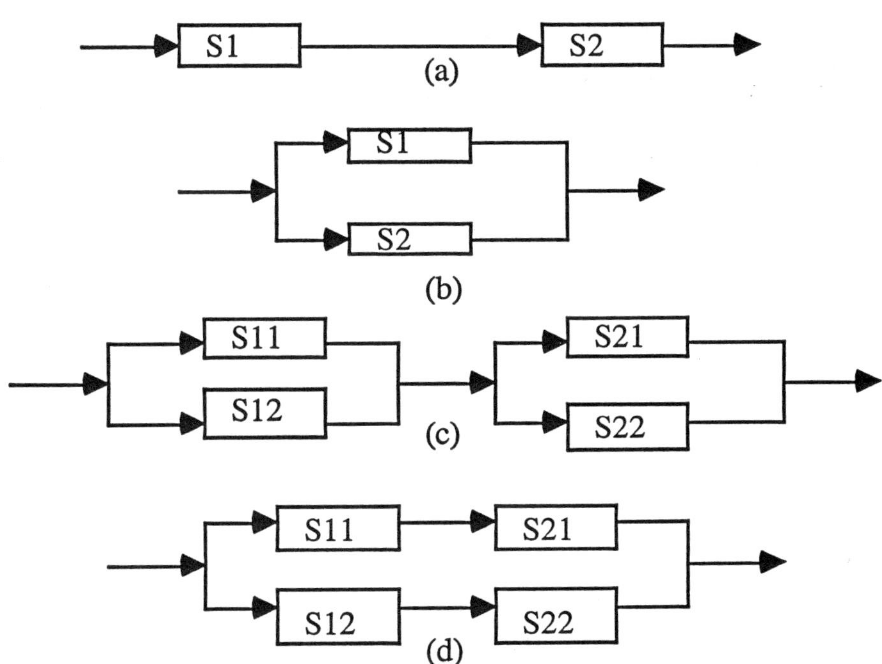

Figure 3.10 Suite of sensors (adapted from Dasarthy). Sensors can be placed in different configurations: (a) serial; (b) parallel; (c) serial/parallel; (d) parallel/serial. The serial configuration is also known as tandem.

3.5   Chapter Summary

There are two primary aspects to sensors: concrete and abstract. Concrete sensors have attributes, which must be taken into account in the design of sensor processing systems. Abstract sensors allow sensor readings to be independent of the technology used; they are essential in combining competitive as well as complementary data.

Sensor data can be competitive, complementary, cooperative, or independent. It is important to combine sensor readings in a meaningful manner. The two most important cases are competitive sensor integration and complementary sensor integration.

3.6   References and Further Reading

References to some animal senses as motivation for sensor processing can be found in [50] and [5]. For material on machine vision, see for example [3, 39, 40].

The terms concrete sensor and abstract sensor were first presented in [53]. They are also elaborated in [35, 60].

For a detailed discussion of some commonly used sensors, see [69]. The limitations of single sensor systems have been pointed out by many researchers, including [1, 8]).

N-modular redundancy was popularized by von Neumann [79].

Sensor integration may be viewed as a form of distributed problem solving. By this we mean a cooperative solution of problems by a decentralized and loosely coupled collection of knowledge sources and sensors.

For distributed decision fusion, refer to the papers by Li and Sethi [48, 49].

Problem Set 3

3-1. Animals possess a myriad of senses. Give some examples. Have they provided motivation for the development of intelligent sensor-based systems?

3-2. Is the brain an inspiration for computing and communication paradigms?

3-3. What is a feedback control system? Describe two feedback control systems that occur in nature.

3-4. Attempts to understand and/or mimic human behavior have resulted in interesting machine systems and associated research. Can you name some?

3-5. Give a macro/micro view of the human eye in the context of distributed sensor processing.

3-6. Give examples of NMR (N-modular redundancy).

3-7. How many senses are used by a person when he eats? Identify the types of sensor fusion that occur.

Chapter 4

Algorithms for Sensor Integration

The evolution of sensor systems has seen the following: single-sensor systems, replicated sensors, disparate sensors, spatially distributed sensors, and intelligent sensors. A distributed sensor network employs a variety of sensors; the use of diverse, intelligent sensors that are spatially distributed offers several advantages that cannot be achieved by single-sensor systems. The use of multiple sensors calls for a meaningful way to combine the information provided by the individual sensors; the process is referred to as *sensor integration*. Modern sensor processing systems (e.g., detection and tracking systems and, multisensor data acquisition systems) use a combination of replicated and disparate sensors.

Sensors may be replicated physically or logically. Physically replicated sensors are of the same device type with the same physical characteristics. Logically replicated sensors perform the same function although they may be different physically. By choosing an appropriate abstraction, we can deal with data from both physically and logically replicated sensors in a consistent manner; they provide nearly identical and hence redundant data. We will present techniques to combine the redundant abstract sensor readings data in a fault-tolerant manner. The integrated output provides increased reliability over individual sensor readings. Redundant sensor data are also called competitive data; combining them is known as competitive sensor integration.

Disparate sensors can either give independent data or complementary data. By complementary data, we mean partial but useful information about the environment; the partial information is usually combined into global information. Often, there is some overlapping information can be used as reference for registration (or alignment) of the partial information. The process of combining complementary sensor data is called *complementary sensor integration*.

There is no general agreement on the semantics of the terms integration, fusion, competitive, and complementary. We will give definitions, contexts, and illustrative examples to make clear what we are discussing. Care should, however, be taken in reading the literature provided at the end of each chapter and, collectively, at the end of the book.

A system is said to be fault tolerant if its performance is not affected by faults in the system. We use the term fault tolerant sensor integration to refer to techniques of combining sensor data when some of the sensors may be faulty.

Mostly, we will deal with fault-tolerant sensor integration of redundant information provided by the physically and/or logically replicated sensors. For convenience, we will classify our sensor integration algorithms into three groups: **One-dimensional sensor integration:** We will present four methods for integrating one-dimensional data from redundant sensor arrays:

- Algorithm M1 (by Marzullo)

- Algorithm J1 (by Jayasimha)

- Algorithm PIKM (by Prasad, Iyengar, Kashyap, and Madan)

- Algorithm PIKR (by Prasad, Iyengar, Kashyap, and Rao)

Multidimensional sensor integration: We will present methods for integrating data from redundant sensor arrays measuring multiple dimensions. They include the following:

- Algorithm CM (by Chew and Marzullo)

- Algorithm BI (by Brooks and Iyengar)

Hierarchical sensor integration: We will present methods for integrating data from distributed sensors that are arranged as a network with two or more levels.

4.1 Concrete and Abstract Sensors

A sensor is a device that senses a physical variable in the environment. A sensor is usually called a transducer if it transforms the measurement of physical variables into electrical signals. Sensors are commonly employed in control systems surveillance systems and data acquisition systems.

Several researchers have stated the need to differentiate two aspects of a sensor: (1) How is it made up? (2) What is it intended to do?

To describe the first aspect of sensors, the qualifiers *concrete* and *physical* are used. To describe the second aspect of sensors, the qualifiers *abstract* and *logical* are used.

4.1.1 Concrete Sensor

A *concrete sensor* is a device makes observations of the environment. It is also called a physical sensor, since it samples the physical state variable of interest.

A *physical variable* is any measurable feature of the environment.

A *physical value* is the actual value of some physical variable at a particular instant in time. We denote an actual physical value by a function $F^*(t)$.

A thermometer is a concrete sensor. It is used to sense the temperature of the environment. Temperature is a physical variable. Temperature readings are generally given in degrees Fahrenheit, degrees Celsius, or kelvins.

Sensors differ in their attributes (e.g., operating range, precision, technology, or intended application). There are a variety of thermometers in use with different attributes, but the bottom line is that they all sense temperature. It is possible to read the temperature using a pressure gauge; the general gas equation, $PV = nRT$, suggests a relationship between the temperature T of a gas to its pressure P and volume V.

Example 1

A nurse uses a clinical thermometer to monitor a patient's temperature. Does she need to know how a thermometer works to use it? With a digital readout, all she need to do is to decide if the temperature reading is reasonable. If she doubts the validity of a reading, she can cross-check using another thermometer. This suggests that the thermometer may be treated as a black box by its user. A micro view of a thermometer (e.g. How is it constructed? How does it work?) is not needed by a user. A macro view (e.g., What is its function?) is probably sufficient. Thus, in our subsequent discussion, we will deal with abstract sensors.

Example 2

There are several choices in the design of a surveillance system. We can use microwave radar, laser radar, or an infrared sensor. We can run the system in synchronous or asynchronous mode. In synchronous mode, the computer can poll (or sample) the sensor for its reading. In an asynchronous mode, the computer can have an event handler; an exception condition (e.g., detection of a certain signature) will alert the computer.

How should we represent the sensor readings inside a computer? A choice would be to represent a sensor output as a single number. However, there are several reasons why a single number is not adequate.

First, every sensor has limited accuracy due to manufacturing tolerance, aging, environment, and other causes. Second, the presence of noise signals often causes an uncertainty in the output values. Third, there is a communication delay between the time t_s that the value is read and the time t_p that the value is processed by the computer (or a processing element). In distributed systems, it is hard to know the correct values of t_s and t_p, or even reasonable estimates.

To allow for the tolerance of the reading, a range of values, for example, *low ... high*, is preferred.

4.1.2   Abstract Sensor

An *abstract sensor* is a piecewise continuous function that maps a physical variable into a dense interval [*low*, *high*] of physical values. This interval contains the value of the physical variable as read by the physical sensor. We denote an abstract sensor as a function $F(t)$, which is an approximation of the actual physical value $F^*(t)$.

We can also view an abstract sensor as a sensor that reads a physical parameter and gives out an abstract interval-estimate I_s, which is a bounded and connected subset of the real line $\mathcal{R}$. An *abstract estimate* is the interval produced by an abstract sensor. An abstract estimate can be (1) an abstract sensor, (2) the output obtained by combining or fusing abstract estimates. We will use the terms sensor, abstract sensor, and abstract sensor estimate interchangeably, unless stated otherwise.

Note that a concrete sensor is a physical entity and an abstract sensor is a mathematical abstraction. An abstract estimate is useful and convenient in combining different sensor readings.

4.1.3   Relating Concrete and Abstract Sensors

How can we relate the two kinds of sensors? There are two scenarios: (1) concrete to abstract and (2) abstract to concrete.

Concrete to Abstract

We start with a concrete sensor and then derive an appropriate abstract sensor. Suppose a given concrete sensor s reads a value, say v, and its maximum error or tolerance can be Δv. We could then define the abstract sensor to be the interval $[v - \Delta v \ldots v + \Delta v)]$. This is called the interval estimate of the sensor's value.

Abstract to Concrete

We start with the abstract sensor and then try to realize a concrete sensor. This case corresponds to a step in Marzullo's design methodology [13, 53] in which a *process control program* is first formally specified and verified before choosing suitable implementations.

The specification will consist of abstract sensors. The problem then requires finding the corresponding concrete sensor. Determining a function that maps a concrete sensor to an abstract sensor depends on many factors, such as the choice of a particular sensor type (e.g. motion detecting sensor, range finding sensor, or vision sensor), the compensation that has to be applied to the raw sensor value, which is itself dependent on the local values of certain parameters (e.g., design parameters of the sensor), the nature of the application, and others [53].

4.2 Faulty, Correct, and Reliable Sensors

A sensor is said to be *correct* if its interval estimate contains the actual value of the physical variable being measured; otherwise, it is said to be *faulty*. Suppose a sensor reading is given by $[low, high]$, where $low \leq high$. Then the *width* of the interval is given by $high - low$; ideally, the width should be zero. In real-life, faults can be intermittent or permanent. For our discussion, we assume that once a fault occurs, it stays permanent.

There may or may not be a bound on the number of faulty sensors. The bound, if it exists, may or may not be a function of the total number of sensors. Can we further qualify faulty sensors? Mechanical vibrations often cause the needle of a dial or meter to settle in a region that does not contain the correct value but lies close to it. Is a miss as good as a mile? We do not think so in the context of sensor integration; we prefer to treat such faults as tame.

A sensor is said to be *tamely faulty* if it is not correct, but its output overlaps with that of a correct sensor. A faulty sensor that is not tamely faulty is said to be *wildly faulty*. For wild faults, there is no correlation between the physical value being measured and the interval estimate of the sensor.

A *reliable* sensor estimate is one that always contains the physical value of interest. The estimate is usually a single contiguous interval; it might however consist of a set of intervals.

4.3 Types of Sensor Integration

There are several ways to combine information from the various sensor outputs. If the sensor provides nearly identical and redundant information, it is called competitive integration. If the sensor provides complementary (or slightly overlapping) information, it is called complementary information. Figure. 4.1 gives one classification of sensor integration.

Competitive Integration

Competitive data are usually provided by replicated sensors sensing the same physical environment. Each sensor ideally provides identical information. Statistics tell us that there will be variations in the information provided by replicated sensors; Some possible reasons include the following:

- Noise at a particular sensor site

- Unreliable transmission due to channel failures

- Loss of accuracy in translating information from the physical readings into a common abstract information medium

Suppose we take three photographs of an object from the same viewpoint. The photographs essentially convey the same information and are redundant. We are

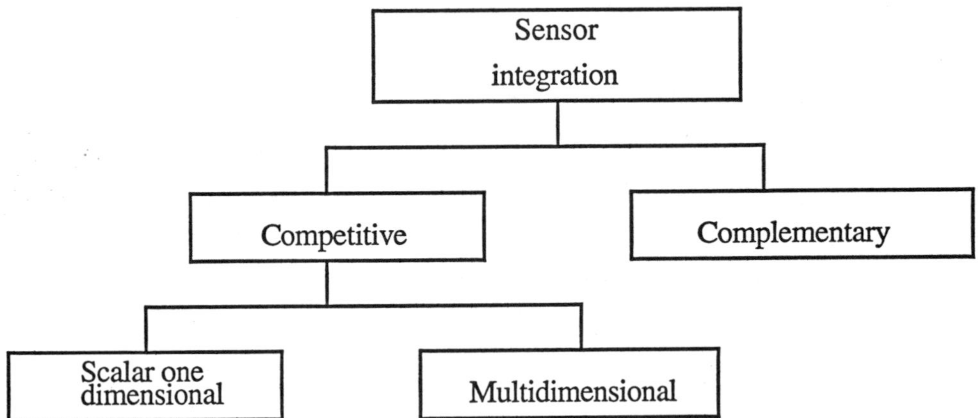

Figure 4.1 Classification of sensor integration. Competitive integration involves replicated sensors. Complementary integration involves disparate sensors. one-demensional sensor readings can be modeled by a scalar value. Multidimensional sensor readings have more than one dimension and can be modeled by a vector.

fairly sure we will not completely lose information even if one photograph turns out to be bad.

Figure. 4.2 gives a classification of competitive scalar integration algorithms. We will describe them later in detail.

In sum, a reason for performing competitive sensor integration is to improve the reliability and fault tolerance of the multisensor data acquisition system.

Complementary Integration

Suppose that only partial information about the environment is returned by each sensor. It could be that no sensor is capable of covering the whole environment (as in multi-target tracking and detection systems). It could also be that the disparate sensors offer different kinds of trade off. (For example, a visible-band satellite gives more accurate and detailed results than infrared sensors, but is more susceptible to atmospheric interference [37].

What is a meaningful way to integrate the disparate sensors to obtain the global information? Suppose we take a picture of a scene from multiple viewpoints at the same time. The photographs have some overlap. By properly piecing the photographs together, we get an overall view of the scene.

Mobile robot systems use a variety of sensors: contact, sonar, infrared, acoustic,

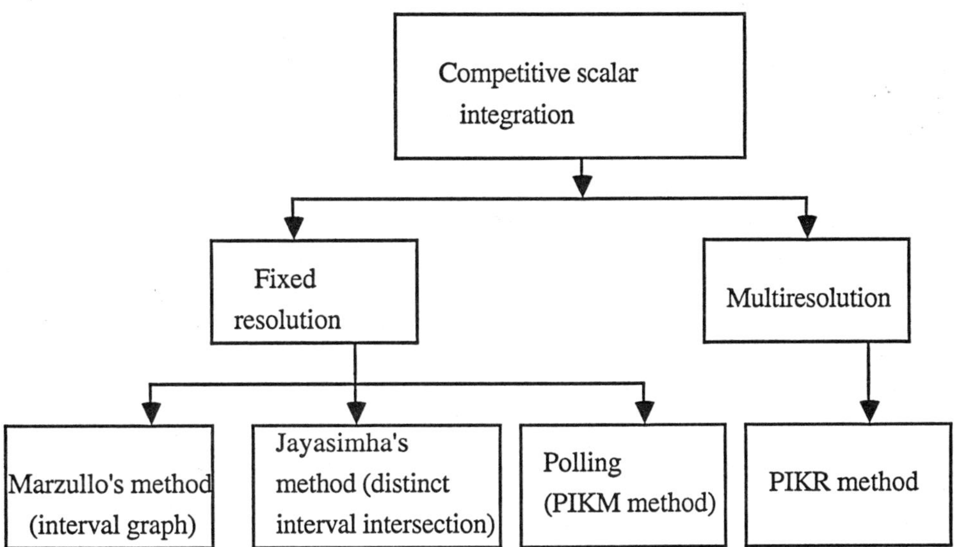

Figure 4.2 Competitive scalar integration algorithms.

stereo vision and laser range finding. Then we need complementary integration.

The term multisensor integration is also used in the literature. It implies that there are multiple sensors and that they are generally disparate. However, we can have both replicated sensors and disparate sensors in a multisensor data acquisition system.

For further reading see [17, 26, 35, 51, 52, 53, 82].

4.4 Marzullo's Method

Marzullo's methodology to specify a fault-tolerant process control system is first described in Marzullo [53] and extended in Chew and Marzullo [13]. The step, which is relevant to our discussion, requires the averaging of N abstract sensors of which F sensors (usually expressed as a function of N) are to be done in a fault-tolerant manner.

Marzullo made clear the need to distinguish between a concrete sensor and an abstract sensor. He reasons out clearly why an abstraction of a sensor reading should be an interval of dense real numbers.

He introduces a model for use in sensor integration. In his model, there are N one-dimensional abstract sensors of which at most F can fail. Assuming F to be a function of N, he determines under what conditions the given sensor readings can be integrated in a fault-tolerant manner.

Marzullo describes two algorithms. For convenience we will refer to them as algorithms M1 and M2. Algorithm M1 is for fault-tolerant integration of N scalar abstract sensors where at most F of them may be faulty. Algorithm M2 is for detection of failed sensors.

4.4.1 Abstract Sensors and Interval Graphs

An abstract sensor estimate S is given as an interval (l, u), where l and u are the lower and upper limits of the interval.

If there are several abstract sensors, say N, we will write $S_1, S_2, ..., S_N$. The ith sensor estimate S_i is then given by (l_i, r_i).

We define an ordering on sensor estimates as follows: $S_i \leq \cdot\, S_j$ if (1) $l_i < l_j$ or (2) $l_i = l_j$ and $u_i \leq u_j$.

The ordering can be used to sort N given redundant sensor estimates in $\Theta(N \log N)$ time.

We can represent a collection of N competitive abstract sensors as an interval graph. Marzullo [53] has shown that abstract sensors are isomorphic to interval graphs.

Given N abstract sensors (each expressed as an interval), can construct an interval graph by associating each interval with a vertex and joining the two vertices if and only if the corresponding intervals intersect. For details, refer to a graph theory text such as Golumbic [23].

We are now ready to discuss Marzullo's algorithm for fault-tolerant integration of redundant abstract sensors.

4.4.2 Algorithm M1

The inputs to Marzullo's algorithm M1 for fault-tolerant sensor integration are as follows:

- N, total number of abstract sensors

- A set of N abstract sensor estimates, say $S_1, S_2, ..., S_N$

- F, maximum number of faulty sensors

The output of algorithm M1 is I_s, which is defined to be the smallest interval that contains the actual value of the physical variable. Then the algorithm can be described in pseudocode as follows.

Algorithm :

begin

    Step 1: Sort the N given sensor inputs such that
        the lower bounds are non-decreasing.

    Step 2: Construct the corresponding interval graph G.

    Step 3: Find all possible $(N - F)$-cliques in the interval graph G.

    Step 4: Find the smallest safe interval that contains
        all the possible $(N - F)$-cliques.

end

The sensor averaging algorithm is based on masking failures through redundancy [72]. It is a generalization of N-moduleredundancy (NMR), whereby N independent copies are fed into a majority voter [79]. The algorithm has a complexity of $\Theta(N \log N)$.

Marzullo finds all the $(N - F)$ cliques in the interval graph. This is equivalent to finding all $(N - F)$-intersections of sensor interval estimates. The reasoning behind this step is as follows.

By definition, any correct sensor must contain the physical value (i.e., the actual value of the physical variable in the environment). We make a key observation: any two correct sensors must overlap since they both contain the actual physical value. We are given that at most F of the N given sensors can be faulty. Thus, at least $(N-F)$ sensors are correct and hence their intervals must contain a common overlap region. A sensor that does not contribute to an $(N-F)$-overlap is necessarily faulty.

Only one of the $(N - F)$-intersections of the N sensors contains the value of the parameter sampled. It is not generally possible to decide which of the $(N - F)$-overlaps contains the parameter's value. Thus, choose and return a smallest safe result.

Let l be the smallest value contained in at least $(N - F)$ of the intervals in S, and let h be the largest value contained in at least $(N - F)$ of the intervals in S; then, the *smallest safe* interval $I_{(F,N)}(S)$ is given by the interval $[l, h]$ [53].

4.4.3 Example

We will illustrate Marzullo's method with two examples.

In Example 1, we are given five abstract sensor estimates. Each sensor estimate is represented by an interval, with its lower bound and upper bound.

We construct an interval graph. The vertices of the interval graph are labeled and correspond to the abstract sensor estimate. An edge connects two vertices if and only if the corresponding intervals have a common intersection.

Figure. 4.3 shows a set of N, say 5, intervals with their upper and lower bounds and the corresponding interval graph. Assume that at most F, say 2, sensors can

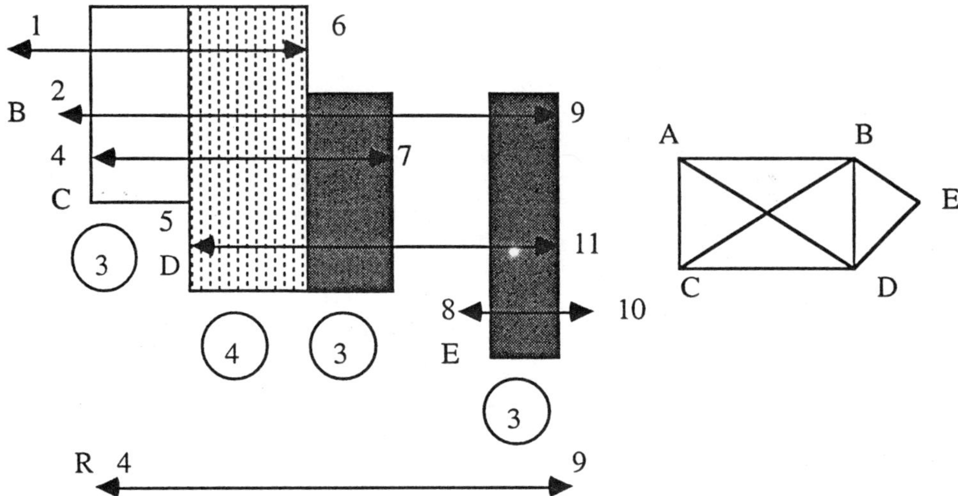

The one interval reliable abstract sensor

Figure 4.3 An interval graph of five abstract sensor estimates.

be faulty. Algorithm M1 finds all the $(N - F)$-cliques in the interval graph. Clearly the cover of all the $(N - F)$ cliques contains the correct value. In our example, all 3-cliques are examined. Their union definitely contains the actual value.

Example 2 considers a case where there are $N = 6$ sensors and at most $F = 3$ can be faulty. We do not show the interval graph explicitly in this example. We consider all $(N - F) = 3$ intersections as shown in Fig. 4.4.

4.4.4   Observations

Algorithm M1 produces a *safe* but large result, as shown in the figure. The width of the integrated output reflects the accuracy of sensor integration; ideally, it should be zero.

Why does Marzullo choose to construct the reliable abstract sensor as a single correct interval of bounded width? It can keep down the communication requirements, and also it preserves the shape of the abstract sensor.

I_s is defined to be the smallest connected interval containing all the $(N - F)$-intersections. I_s is safe, but far too conservative; it contains the actual physical value, but it may also include points in the correct interval do not belong to any of the intervals. Note that this estimate can become arbitrarily wide as the number of sensors becomes large.

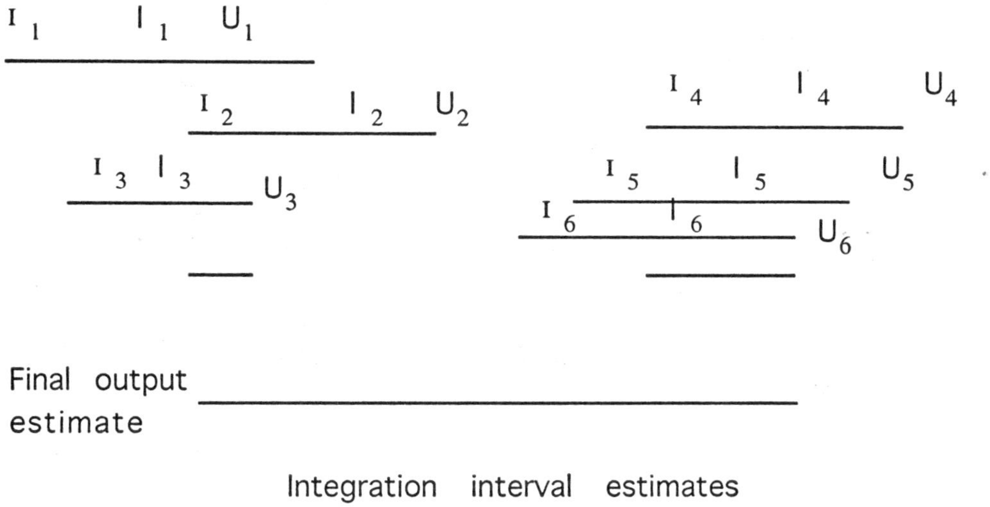

Figure 4.4 A set of six intervals such that $l_1 < l_3 < l_2 < u_3 < u_1 < l_6 < l_5 < l_4 < u_6 < u_5 < u_4$. Of the six sensors, assume that at most three can be faulty.

Algorithm M1 is generalizable to higher dimensions [13].

Marzullo has derived bounds of the output estimate, where F is expressed as a function of N (see Theorems 1 and 2). In particular, Marzullo [53] shows that if the width of the single-interval reliable abstract sensor is to be bounded by the width of a correct abstract sensor then $F < N/3$.

Theorem 1 If $F < \left| \frac{N+1}{2} \right|$, then the width of $I_{[F,N]}(S)$ is bounded from above by the $(2F+1)^t h$ smallest width in S.

Theorem 2 If $F < \left| \frac{N}{3} \right|$, then the width of $I_{[F,N]}(S)$ is bounded from above the $(F+1)$-th smallest width in S.

Algorithm M2 is a variant of M1 and is designed to detect failed sensors. However, algorithm M2 can fail to detect some of the incorrect sensors.

4.5   Jayasimha's Method

Jayasimha assumes essentially the same model proposed by Marzullo. There is a major difference, however. Since at most F out of N sensors can fail, Marzullo tries to find all possible $(N-F)$-intersections. Jayasimha notes that since F is an upper bound for the number of faulty sensors, there could be $(N-F+1)$-intersections and even N-intersections that contain the correct sensor value.

Suppose $N = 5$ and $F = 2$. Marzullo finds all 3-intersections; Jayasimha notes that 4-intersections and 5-intersections could also contain the correct sensor value. Thus, Jayasimha introduces a stricter notion of interval intersection as follows: Given N intervals, are N-distinct $(N-I)$ intersection is an interval in which $(N-I)$ intervals, but no more, intersect.

In this section, the term interval intersection means distinct interval intersection unless otherwise stated.

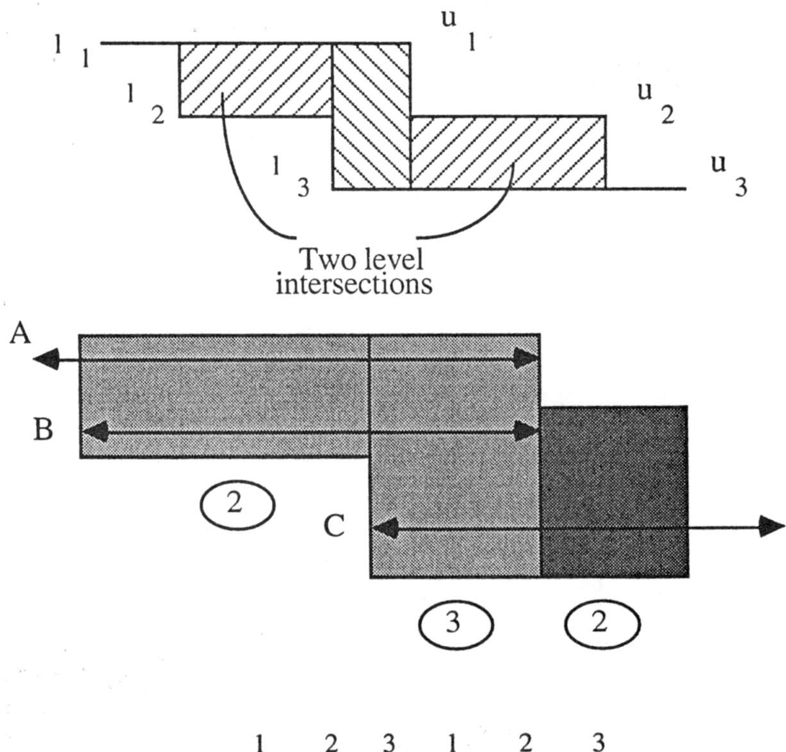

Figure 4.5 Three sensors are given such that $l_1 < l_2 < l_3 < u_1 < u_2 < u_3$. There is a two-level intersection (involving A and B), a three-level intersection (involving A, B and C), and a two-level intersection (involving B and C).

Jayasimha associates each interval with a numeric label that indicates the number of intervals that overlap with it. All the initial intervals (representing the sensor estimates) are assigned the label 1. As algorithm J1 discovers overlapping intervals, it updates the labels of the intervals. If I intervals with label 1 intersect, the result of the intersection of I intervals is assigned the label I.

An interval may be broken up into several intervals by the algorithm J1. Fig. 4.5 shows two-level and three-level intersections giving three intervals A, B, and C. In Jayasimha's convention, all intervals, sorted by their lower bounds, are depicted from top to bottom and left to right as in Fig. 4.5. Interval intersections are shown by shaded regions, with a circled number x below to represent an x-interval intersection. In Figure. 4.5, the intervals denoted by the shaded regions form a contiguous interval.

Since we are given N independent abstract sensor estimates of which at most F can be faulty, it is obvious that any point not contained in at least $(N - F)$ intervals cannot be the correct value. Algorithm J1 constructs the reliable estimate as a collection of the minimum number of $(N - F)$ or more intersecting intervals one of which contains the correct value. The single correct interval (as given by Marzullo) can be obtained by taking the union of all the intervals that arise in his method.

4.5.1 Algorithm J1

The inputs to the algorithm are N, the total number of sensor estimates; N independent sensor estimates $S_1, S_2, ..., S_N$; and F, the maximum number of faulty sensors.

The output of algorithm J1 is a list containing all the regions where $(N - F)$ or more sensors intersect. For the one-dimensional case, a region will be an interval. The output will be a list of intervals.

We can easily modify algorithm J1 to output the the set of members, and also the set of nonmembers associated with the list.

The algorithm is as follows:

Algorithm
begin
    Step 1: Sort the N given sensor inputs such that
        the lower bounds are nondecreasing.

    Step 2: Initialize all labels to 1.

    Step 3: Begin with the $(N - F)$-th sorted interval.
        Check if this interval intersects with all the
        (N-F-1) earlier intervals.
        If it does, then record this intersecting interval
        with the label $(N - F)$ in the set I.

    Step 4: Discard the interval.
        Proceed to the next interval and check if this
        interval intersects with all the $(N - F - 1)$ earlier
        intervals, excluding the discarded interval.

> Record the intersecting interval and check the set I
> to see if this new intersecting interval
> intersects with other interval(s) in I.

Step 5: Repeat step (4) until the last interval has been examined.

Step 6: Output all labels with $(N - F)$ or higher,
and the members that belong to the cliques.
Also, output the single contiguous interval
that contains the actual value.

end

A more detailed version of the algorithm is given in [34]. It also gives a proof of correctness. Another version is given in [9]; the modified J1 algorithm is used as a procedure for multidimensional sensor integration.

4.5.2 Implementation Choices

We have given a high-level description of algorithm J1 in pseudocode. There are several possible design and implementation choices.

First, we could represent an abstract sensor estimate S_i as a triplet $\langle id, l_i, u_i \rangle$, where

- id stands for a unique sensor identification number ID in the range $1 \ldots N$.

- l_i is the lower limit of the ith reading.

- u_i is the corresponding upper limit.

For the one-dimensional case, l_i and u_i will be both scalars. For the multidimensional case, l_i and u_i will be vectors.

Second, we could maintain the S_i's as a sorted list. This makes the proof of the algorithm easy. We could use the plane sweep technique from computational geometry.

4.5.3 Example

Figure. 4.6 presents a line that represents the domain mapping a set of continuous sensor readings. Each sensor provides its data as an upper and lower bound. Data from seven sensors are represented in Fig. 4.6 by lines above the axis.

Redundancy is provided for at most F sensors, where F is less than $N/2$. So, from the seven sensors, we can tolerate failure in at most three. If up to three readings can be false, this means that the actual value must be contained in a region where the readings from $(N - F)$ of the sensors intersect. Figure. 4.6 shows the four regions for the example data beneath the axis.

The ranges shown above the axis are ranges read on sensors. The areas below
the axis are intersections of N-F of the sensors. The mentional case with
seven sensors (N=7) upto three of which can be faulty (F = 3).

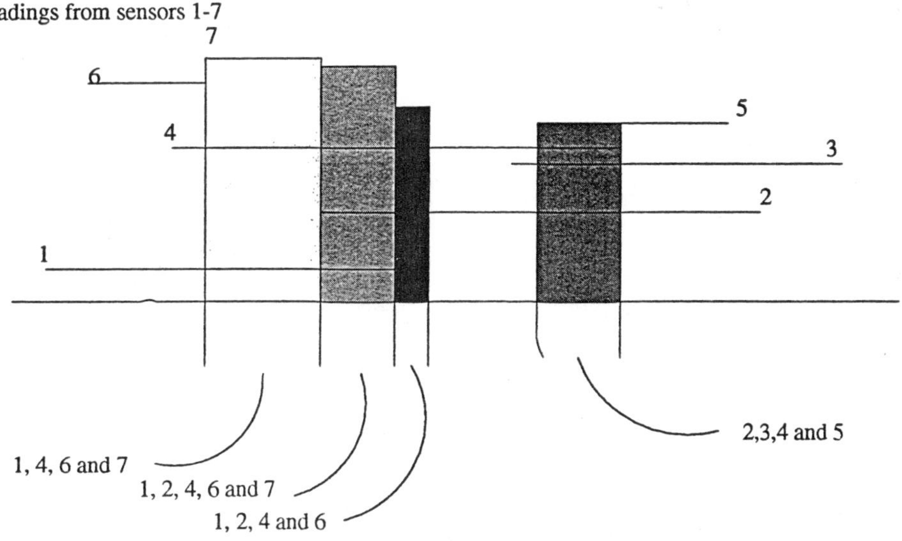

Figure 4.6 Example one-dimensional sensor integration.

We do not know which of the regions in Fig. 4.6 contain the value of the physical
variable. For this reason, algorithm J1 returns a region defined by the smallest lower
bound and the largest upper bound.

Along with the one dimensional region, Algorithm J1 returns a list of all regions
where $(N - F)$ sensors agree. It provides a list of possible combinations of faulty
sensors.

Algorithm J1 finds four cliques: $\{1, 4, 6, 7\}$, $\{1, 2, 4, 6, 7\}$, $\{1, 2, 4, 6\}$, and

$\{2, 3, 4, 5\}$. The elements excluded from these cliques are possibly faulty. In our example, they are $\{2, 3, 5\}$, $\{3, 5\}$, $\{3, 5, 7\}$, and $\{1, 6, 7\}$.

Algorithm J1 sorts the sensor readings based on their upper and lower bounds. Since these are dominant operations in the algorithm, J1 has a complexity of $O(N \log N)$.

How does J1 compare with M1 and M2?

- M1 uses interval graphs. J1 does not use interval graphs; it recognizes all the maximal cliques of size $(N - F)$ or more.

- M1 finds all possible $(N - F)$ intersecting intervals and returns I_s which is the smallest contiguous interval containing the actual value. J1 finds all the $(N - i)$ $(0 \le i \le F)$ intersecting intervals and the participating intervals.

- M1 and J1 both have $\Theta(N \log N)$ complexity and are optimal. However, J1 has possibly better runtime constants than M1.

- M2 can fail to detect some faulty sensors. J1 can detect all the possible sets of faulty sensors.

- J1 has tight upper bounds on the number of $(N - F)$ intersecting intervals and the total number of $(N - i)$ $(0 \le i \le F)$ intervals. They are $F + 1$ and $2F + 1$, respectively.

Consider the case of five redundant sensors of which two may be faulty (see Fig. 4.3). Here, $N = 5$, $F = 2$, and $N - F = 3$.

For the example, algorithm M1 finds all the 3-cliques in the interval graph. Clearly, the cover of all the 3-cliques contains the correct value. Note that the 3-cliques ABD and ACD are recognized even though they yield the same 3-interval; they are subsumed by the 4-clique $ABCD$.

Algorithm J1 does not explicitly use interval graphs. It recognizes all the maximal cliques of size $(N - F)$ or more. For the example, it recognizes three maximal 3-cliques ABC, BCD and BDE and one maximal 4-clique $ABCD$. The single-interval reliable abstract sensor is denoted by R; it is returned by both M1 and J1. The actual reliable abstract sensor is the set of four line segments represented by the shaded regions; it is returned only by J1.

A bound can be placed on the number of intersecting intervals. The total number of $(N - F)$ or more interval intersections in a set of N intervals with at most F of them faulty is at most $(2F + 1)$. The number of $(N - F)$ interval intersections in a set of N intervals with at most F of them faulty is at most $(F + 1)^2$. For proof, see [34]

4.6 PIKM Method

We will describe the Algorithm PIKM proposed by Prasad, Iyengar, Kashyap and Madan [60]. PIKM follows Marzullo's model, but with a difference. Like M1, PIKM assumes that N redundant sensors are given and that at most F can be faulty. M1 assumes that F is a function of N; PIKM does not require F to be a function of N. M1 assumes that all faults are equally likely. PIKM assumes that all faults are not the same; some could be tamely faulty while others could be wildly faulty. A tame fault is one that has an overlap with a correct sensor. By itself, a tame fault does not provide the correct answer; however, when the number of sensors is large and most faults are tame, the clustering effect of the tamely faulty sensors should become prominent. This heuristics is at the heart of the PIKM algorithm. Experimental validation is done with the use of a simulator program.

The PIKM model for sensor integration can be summarized as follows:

- Given N linear redundant sensors.

- N is large.

- At most F sensors can be faulty.

- Most faults are tame.

Like M1, PIKM represents an abstract sensor as a dense interval on a real number line; it does not care where the interval lies or how wide it is.

As an aside, PIKM views an abstract sensor as a set with an associated characteristic function that determines its membership. This is the first and important step in formalizing the sensor algorithms used by PIKM and M1.

Let us review the example discussed earlier (see Fig. 4.4). We are given a set of six intervals where $l_1 < l_3 < l_2 < u_3 < u_1 < l_6 < l_5 < l_4 < u_6 < u_5 < u_4$. Assume that at most three can be faulty. Algorithm M1 produces a safe but large result by considering all $(N - F)$-cliques.

The width of the integrated output reflects the accuracy of sensor integration; ideally, it should be zero. To narrow the width (in the case of ties), we need additional information and/or assumptions. For instance, we may define a *popularity* or equivalent measure.

Under the PIKM model, the sensor integration algorithm can exploit the clustering effects of the tamely faulty sensors to suggest potential areas where the correct value might lie. Algorithm PIKM can substantially reduce the width of the output interval estimate in more cases than algorithm M1.

4.6.1 Algorithm PIMK

Algorithm:
Input: Intervals $I_1, I_2, \ldots, I_n$, and f (parameter representing maximum number of faulty sensors).

Output: Integrated output estimate.

begin

1. Take all $(N - F)$-intersections of the intervals to yield intervals $L_1, \ldots, L_k$, each of which is an $(N - F)$-intersection:
   $$\{L_j = [a_j, b_j]\};$$

2. For each i $(1 \leq i \leq k)$

   (a) Count the number of intervals intersecting each of the intervals $I_j (1 \leq j \leq n)$ having nonempty intersection with L_i

   (b) Add these numbers to obtain a number r_i. { r_i gives the sum of the number of intervals intersecting with intervals involved in the formation of the L_i. a is a measure of the reliability of L_i}

3. Choose the maximum of the $r_i (1 \leq i \leq k)$ and call it r

4. Let $m = \min\{i | r_i = r\}$ and $M = \max\{i | r_i = r\}$

5. Assign $I_p^* = [a_m, b_M]$ to be the integrated output estimate

end.

Table 4.1 Popularities of Intervals

Interval	I_1	I_2	I_3	I_4	I_5	I_6	I_7	I_8	I_9	I_{10}	I_{11}	I_{12}	I_{13}
Popularity	4	2	4	4	3	2	4	4	4	2	2	4	3

In the PIKM model, the $(N - F)$ intersections are assigned weights. In the example, the weights represent reliability estimates of these intersections; we may impose any convenient rule for choosing these segments according to their reliabilities.

Table 4.1 shows each interval with its popularity. The $(N - F)$ or more intersections (i.e., 3 to 4 intersections here) that form the output have reliabilities 8, 12, 15, 15, 10.

In the PIKM method, the intersection is weighted to help us judiciously choose the output intervals. We may employ any convenient rule depending upon our faith in the tameness of the faults to pick these intervals and enclose them by a connected interval. For instance, we may choose only those intervals with maximum reliability (in this case, the intervals with reliability 15) and enclose them by a connected interval. It is clear that the worst possible width for the final output interval estimate is the smallest interval containing all intersections irrespective of their weights.

To address the general problem of *fault-tolerant* sensor integration for a large class of sensors, it is necessary to evolve a broad-based computational framework that can accommodate a wide range of sensors and a variety of fault-tolerant integration techniques depending on the phenomenon being sensed and the method of sensing.

4.6.2 Observation

The PIKM method may be summarized as follows:

- A *calculus* of sensor integration is developed treating the sensor estimates as subsets of an abstract parameter space and obtaining functional representations of the characteristics of these estimates. The basic functions can then be combined to get functions describing the characteristics of the output according to the kind of integration that is required to be performed.

- The fault-tolerant integration of abstract interval considers a new failure model wherein we can reduce the width of the output interval estimate significantly in cases involving a large number of sensors.

- The algorithm assumes that *weights* are assigned to the intervals and computes easily the interval(s) that is(are) *most popular*. The algorithm is guaranteed to give results that are not worse than those given by Marzullo's method.

    1. When sensors with tame faults cluster prominently around the correct sensors, PIKM results are better than Marzullo's.

    2. If the effect of tame faults is not prominent, PIKM results are the same as Marzullo's.

- The algorithm has a $\Theta(N \log N)$ complexity. It is also easy to implement the algorithm efficiently.

- The method is not robust against noise.

- There is no trade-off between computation time and output quality.

4.7 Chapter Summary

We have described three techniques for integrating redundant one-dimensional abstract sensors. The first is by Marzullo, who builds an interval graph from the given sensors and then finds all possible $(N - F)$-cliques, where N is the total number of sensors and F is the number of sensors that can be faulty. He gives a safe estimate, which is the smallest contiguous interval containing the actual value.

The second is by Jayasimha, who instead of building an interval graph, maintains sorted lists of the intervals. He introduces the notion of strict interval intersection and associates with each interval intersection the number of intervals participating in the intersection (and also their identities). In this way, the set of intersecting intervals that can possibly contain the actual value is reported. The union of such intervals will give the same output estimate as given by Marzullo.

The third techinque is by Prasad, Iyengar, Kashyap, and Madan. They assume that the number of sensors is large and that most faults are tame. In theory, the clustering effect of the tame faults should occur near the interval intersections that contain the actual value. Popularity and reliability measures are used to select those intervals where the clustering effect is prominent. In essence, PIKM uses a tie breaker to narrow the choices of the intersecting intervals.

4.8   References and Further Reading

Marzullo's work appears in [53] and has been extended in [13]. Jayasimha's work is reported in [34] and has been extended in Brooks and Iyengar [9, 90]. The PIKM algorithm was first proposed in [60].

Problem Set 4

4-1. What is a process control system?

4-2. Give an example of a process control system. What is the relevance of sensor integration in a process control system?

4-3. Suppose we have two sensors S_1 and S_2. At time t, their readings are $S_1 = [l_1, u_1]$ and $S_2 = [l_2, u_2]$. How can we integrate the readings if the two sensors overlap?

4-4. For Problem 4-3, assume that $l_1 = 1.3$, $l_2 = 1.4$, $u_1 = 1.5$ and $u_2 = 1.6$. Find the width of the integrated output.

4-5. Repeat Problem 4-3 for the case where the two sensor readings barely overlap. If there are slight perturbations in the sensor readings, would the integrated output remain the same?

4-6. Draw interval graphs for the case where there are two, three and four sensors.

4-7. Given three sensors S_1, S_2 and S_3. At time t, their readings are $S_1 = [l_1, u_1]$, $S_2 = [l_2, u_2]$, and $S_3 = [l_3, u_3]$. How can we integrate the sensor readings if all three sensor readings overlap?

4-8. Repeat Problem 4-7 for the case where two sets of two sensor readings overlap.

4-9. Repeat Problem 4-7 for the case where only one set of two sensor readings overlap.

4-10. Repeat Problem 4-7 for the case where all three sensor readings do not overlap.

4-11. How can the integration of three sensors be applied to an intelligent node in a complete binary tree organization?

4-12. Show that algorithm M1 is correct.

4-13. What is the union of intervals?

4-14. What is a continguous interval?

4-15. Implement algorithm J1 in a Pascal-like language.

Chapter 5

Sensor Integration using Multiresolution Decomposition

Multiresolution is a useful technique that has been developed independently by numerous researchers while working in their own disciplines e.g., physics, engineering, image processing, and signal processing. Recently, this technique has been used by the wavelet community. We will use multiresolution decomposition as a robust technique to integrate sensor readings.

5.1 Multiresolution Decomposition

Suppose we are given a map with differing resolutions, and we want to find out the coastline in that map. What should we do? We want the map with the most appropriate resolution that is, one that is not too coarse or too fine. A very fine resolution will give too much detail that we do not need, whereas a very coarse resolution will overlook the features that we want to see. We need the right amount of resolution.

Suppose we start with a map with very fine resolution. If there is too much detail, we can go to a map with a larger scale. Effectively, we zoom out until the scale of the map is suffcient for us to discern the coastline.

Suppose we start with a map with very coarse resolution. If there is too little detail, we can go to a map with a smaller scale. Effectively, we zoom in until the scale of the map is suffcient for us to discern the coastline.

Suppose we have a receiver with band spread capability. We can switch quickly from one frequency band to another frequency band. Once we have selected a frequency band, we can discriminate the stations within the band easily by fine tuning. Instead of a single resolution, the receiver offers us a multiple level of resolutions.

In the context of signal processing, we have the following definitions: Given a sequence of increasing resolutions $\{r_j\}_{j\in Z}$, the details of a function $f(x)$ at the resolution r_j are defined as the difference of information between the approximations of $f(x)$ at the resolution $r_j + 1$ and the approximation at the resolution r_j.

A multiresolution representation also provides a simple hierarchical framework for interpreting the signal content. For instance, in an image, it is hard to recognize that a small object is a particular part or feature of a larger object without recognizing the context of the larger object. It is therefore natural to first analyze image details at a coarse resolution and then increase the resolution.

The approximation of a signal $f(x)$ at a resolution r is defined as an estimate of $f(x)$ derived from r measurements per unit length. These measurements are computed by uniformly sampling at a rate r the function $f(x)$ smoothed by a low-pass filter whose bandwidth is proportional to r. To be consistent when the resolution varies, these low-pass filters are derived from a unique scaling function that is dilated by the resolution factor r

Starting at the coarsest resolution, we select those crests with the highest peaks (wavelet components with the largest amplitude) and choose the crest with the widest spread. At the next higher resolution, this crest is again inspected for crests within it with highest amplitudes and among these crests, the one with the widest spread is retained for similar analysis at the next resolution. This procedure results in isolating those regions of the real line over which $O(x)$ has a maximum value, corresponding to degree of overlap. Figure. 5.5 illustrates this procedure. $O(x)$ in Fig. 5.4 is processed using multiresolution decomposition. The advantage of this procedure is significant from the point of view of computational speed, since the coarse-to-fine processing leads to elimination of large regions of the support of $O(x)$ at each resolution.

5.2   Multiresolution as a Basis for Sensor Integration

The three techniques for fault-tolerant integration of abstract interval estimates, algorithm M1 (by Marzullo), algorithm J1 (by Jayasimha), and Algorithm PIKM (by Prasad, Iyengar, Kashyap, and Madan), are optimal, but they are not suitable for real-time applications.

We will now present algorithm PIKR (by Prasad, Iyengar, Kashyap, and Rao) which is suitable for real-time applications. It uses multiresolution decomposition on the overlap function. We describe the case for integrating abstract one-dimensional sensors although algorithm PIKR can be generalized easily to multiple dimensions.

Like algorithm PIKM, PIKR assumes that a large number of faults are possible, but that most faults are tame. If the number of faults are not strictly bounded, the $(N - F)$-intersections tend to be scattered wildly over the real line, giving poor output estimates.

How can we improve the output estimate? We can further evaluate the $(N - F)$ intersections to choose the best possible intersection that contains the correct value with high reliability.

As the sensors are sampled synchronously at various time intervals, algorithm PIKR does the following:

- Orders the sensors a priori by labeling them

- Dynamically maintains the overlap function $O(x)$ of the sensors (see Figure 5.1).

- Analyzes the overlap function at various scales to obtain successively smaller regions that contain the correct value of the parameter observed.

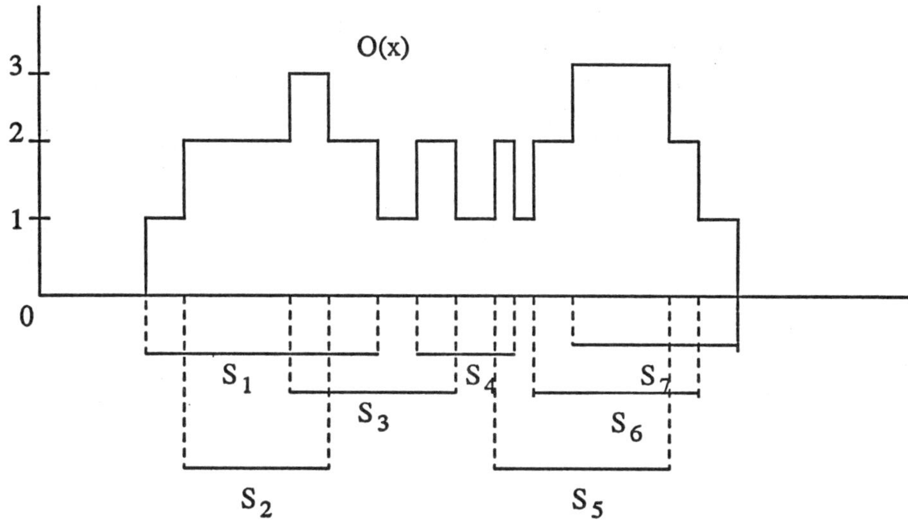

Figure 5.1 Overlap function $O(x)$ for a set of 7 Sensors.

Assume that the length of each interval estimate is bounded below by l and above by L, where $l < L$ and l, L are positive real numbers. Figure. 5.2 describes the regions about the correct parameter value c where the faulty sensors cluster. A very large interval estimate is too inaccurate to be of any value and hence may be discarded. On the other hand, a very small interval estimate would not be amenable for fault-tolerance analysis. A minimum tolerance of $\pm l/2$ is built into the abstract sensors, and so we may assume that the width of each interval is at least l.

Tame faults cluster in a bounded neighborhood around the correct value of the measured parameter. When there is a large number of faulty sensors and most faults are tame, the overlaps of the faulty sensors among themselves can boost the value of $O(x)$ in the neighborhood of the correct value of the parameter, thus reinforcing the $(N - F)$-intersection containing the correct value.

Let T be the number of tamely faulty sensors. These may range in width from l to L. A tamely faulty sensor must intersect with a correct sensor. Therefore its end point nearest to the correct value c must lie within a distance of at most L from c. Thus at most $(1 + \lfloor L/l \rfloor)$ tamely faulty sensors can be accommodated on either side of c with no two of them overlapping. That is, at least $\lceil T/2(1 + \lfloor L/l \rfloor) \rceil$

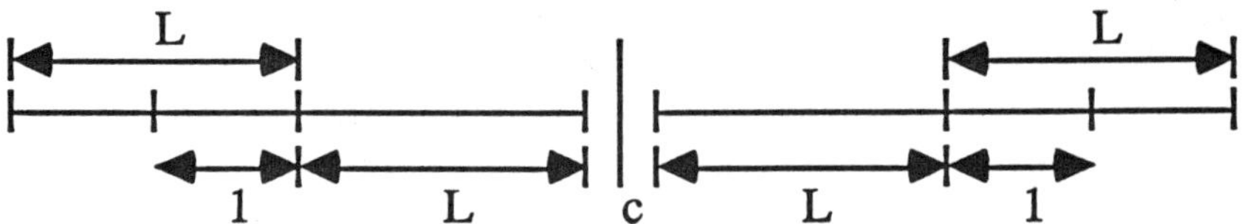

Figure 5.2 Regions about the correct parameter value c where the faulty sensors cluster has $L = 2l$.

tamely faulty sensors overlap over a region of width at least l within a distance of at most $2L$ from c.

When the number of intersections of tamely faulty sensors is $\lceil T/2(1 + \lfloor L/l \rfloor) \rceil$, the width of this intersection is actually at least $2(l + L)$. When the number of intersections is T, this results in a peak with spread of at least l. This clustering reinforces the width and height of the correct $(N - f)$-intersection by adding in its neighborhood a peak of area Tl at least. In general, this results in a taller and wider peak in the neighborhood of c. The wildly faulty sensors, on the other hand, are random in their location on the real line and, being uncorrelated, tend not to cluster in any small neighborhood. Thus, the $(N - f)$-intersections resulting from them have shorter and narrower peaks representing them in $O(x)$.

5.2.1 Multiresolution of the Overlap Function

If S_i $(1 \le i \le N)$ are N abstract sensors with their interval estimates $[a_i, b_i]$ $(1 \le i \le N)$ having characteristic function χ_i $(1 \le i \le N)$ such that

$$\chi_i(x) = \begin{cases} 1 & \text{if } x \in [a_i, b_i] \\ 0 & \text{if } x \notin [a_i, b_i] \end{cases}, \tag{5.1}$$

then the overlap function $O(x)$ of these N sensors is given by $O(x) = \sum_{i=1}^{N} \chi_i(x)$.

For each j, $O(x)$ can be sampled at regular intervals $1/2^j$ to obtain the jth resolution of $O(x)$ at scale $1/2^j$ as a linear combination of a set of functions obtained by scaling and translating a single function.

Let

$$\sigma(x) = \begin{cases} 1 & rmif\ 0 \le x \le 1 \\ 0 & \text{otherwise} \end{cases} \tag{5.2}$$

Figure 5.3 shows depicts such a function.

Let $\alpha \in R$ and $j \in Z$. Consider the functions

$$\{\sigma(2^j(x - \alpha) - n\}_{n=-\infty}^{\infty}\ \sigma(2^j(x-\alpha)-n = \begin{cases} 1 & \text{if } \alpha + n/2^j \le x < \alpha + (n+1)/2^j \\ 0 & \text{otherwise} \end{cases}$$

$$\tag{5.3}$$

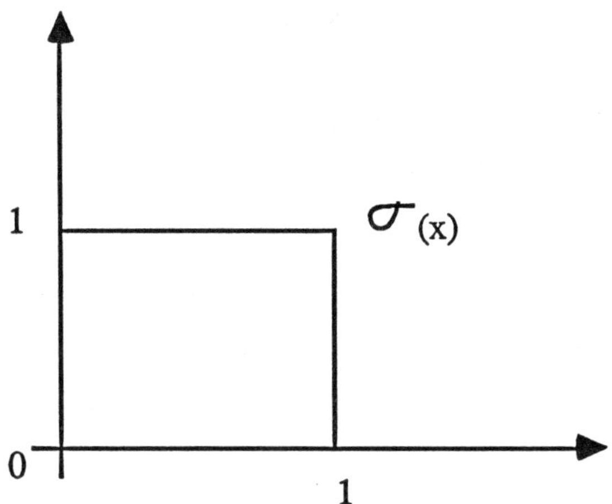

Figure 5.3 Characteristic function of the unit interval $[0, 1]$.

Without loss of generality, we may assume $0 \leq a < 1/2^j$. Note that

$$[\alpha + n/2^j, \ \alpha + (n+1)/2^j) \bigcap [\alpha + (n+1)/2^j, \ \alpha + (n+2)/2^j) = \emptyset \qquad (5.4)$$

and

$$\bigcup_{N=-\infty}^{\infty} [\alpha + N/2^j, \alpha + (N+1)/2^j] = R. \qquad (5.5)$$

The jth resolution of $O(x)$ with respect to the functions $\{\sigma(2^j(x-\alpha)-n\}_{n=-\infty}^{\infty}$ is denoted by $O_\alpha^j(x)$ and is given by

$$O_\alpha^j(x) = \sum_{n=-\infty}^{\infty} O(\alpha + 2^{-j})\sigma(2^j(x-\alpha)-n). \qquad (5.6)$$

Since $O(x)$ has compact support, the above summation is actually over finitely many n. If the interval estimates of the sensors S_i are $[a_i, b_i]$ $(1 \leq i \leq N)$ and $a = \min_{1 \leq i \leq N}\{a_i\}$ and $b = \max_{1 \leq i \leq N}\{b_i\}$, then

$$O_\alpha^j(x) = \sum_{n=\lceil 2^j(a-\alpha)\rceil}^{\lceil 2^j(b-\alpha)\rceil} O(\alpha + n2^{-j})\sigma(2^j(x-\alpha)-n). \qquad (5.7)$$

Thus, $O_\alpha^j(x)$ is obtained from $O(x)$ by sampling $O(x)$ at the points $\{\alpha + n2^{-j}\}_n$ (see Fig. 5.4). O_α^j is a function whose features are of size $1/2^j$ or greater. To study the effect of sampling in the above manner, it is sufficient to study the sampling of the characteristic function of an arbitrary interval $[a, b]$, since $O(x)$ is a linear combination of characteristic functions.

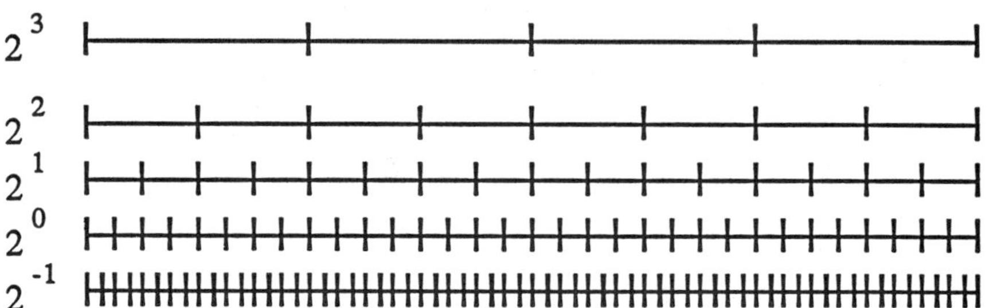

Figure 5.4 Sampling at various scales.

Consider the test function

$$g(x) = \begin{cases} 1 & \text{if } x \in [a, b] \\ 0 & \text{if } x \notin [a, b] \end{cases} \tag{5.8}$$

Case (i) $b - a \geq 1/2^j$, that is, $(g(x)$ is a feature bigger than the scale width. $a \in [\alpha + n/2^j, \alpha + (n + 1)/2^j)$ for some $n, b \in [\alpha + m/2^j, \alpha + (m + 1)/2^j)$, and $n < m$. Thus

$$g_\alpha^j(x) = \sum_{r=-\infty}^{\infty} g(\alpha + r2^{-j})\sigma(2^j(x - \alpha) - r)$$

$$= g(\alpha + n2^{-j})\sigma(2^j(x - \alpha) - n) + \cdots + g(\alpha + m2^{-j})\sigma(2^j(x - \alpha) - m) \text{ if } a = \alpha + n/2^j$$

or

$$g(\alpha + (n+1)2^{-j})\sigma(2^j(x-\alpha) - n - 1) + \cdots + g(\alpha + m2^{-j})\sigma(2^j(x-\alpha) - m) \text{ if } a > \alpha + n/2^j.$$

Therefore we have

$$g_\alpha^j(x) = \begin{cases} 1 & \text{if } \alpha + n/2^j \leq x < \alpha + (n + 1)/2^j \\ 0 & \text{otherwise} \end{cases} \quad \text{if } a = \alpha + n/2^j \tag{5.9}$$

and

$$g_\alpha^j(x) = \begin{cases} 1 & \text{if } \alpha + (n + 1)/2^j \leq x < \alpha + (n + 1)/2^j \\ 0 & \text{otherwise} \end{cases} \quad \text{if } a > \alpha + n/2^j. \tag{5.10}$$

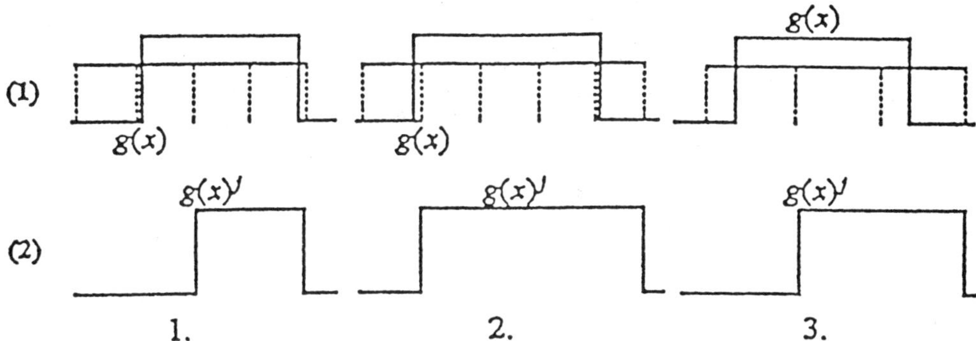

Figure 5.5 Computational characterization of support of $g(x)$.

Thus, two things may happen independently to the support of $g(x)$:
1) It may shrink on the left by at most $1/2^j$.
2) It may extend on the right by at most $1/2^j$ (see Fig. 5.5).

Thus, we need to correct for a positive shrinkage of the support of a feature to avoid loss of information. This is done by resolving over a region bigger than the one at hand by $1/2^j$ on the left. The extension (smearing) of support will decrease with further resolution and does not pose a problem.

Case (ii) $b - a < 1/2^j$. $g(x)$ is a feature smaller than the scale width. If $a, b \in [\alpha + n/2^j, \alpha + (n+1)/2^j)$ for some n, then $g_\alpha^j(x) = 0 \; \forall \; x$; That is, the feature will not appear at scale $1/2^j$. If $a \le 1/2j \le b$ for some n, then

$$g_\alpha^j(x) = \begin{cases} 1 & \text{if } \alpha + n/2^j \le x < \alpha + (n+1)/2^j \\ 0 & \text{otherwise} \end{cases} . \tag{5.11}$$

That is, $g(x)$ will appear as a feature of size $1/2^j$ shifted to the right by at most $b - a$. This will diminish in size with further resolution, and $g(x)$ will be recovered by correction to the left and resolution.

We note that changing a will not produce any advantage insofar as sampling $O(x)$ is concerned, since location of the sampling points $\{\alpha + n/2^j\}$ with respect to $O(x)$ is arbitrary. We may thus conveniently let $\alpha = 0$ and henceforth sample $O(x)$

at points $\{n/2^j\}$ to obtain the jth resolution. Thus,

$$O^j(x) = \sum_{n=\lceil 2^j a\rceil}^{\lceil 2^j b\rceil} O(n2^{-j}\sigma(x2^j - n) \quad j \in Z. \tag{5.12}$$

The fluctuations in $O(x)$ occur at the points a_i, b_i $(1 \le i \le N)$ which are the end points of the interval estimates. If a is the least of the a_i and b is the largest of the b_i, then the average number of fluctuations per unit length is given by $2N/(b-a)$. So, to capture all the fluctuations, we would have to resolve at least to a level $j > \log(2N/(b-a))$.

5.2.2   Selection of Robust Peaks

At the jth level of resolution, O^j_α can be looked on as a series of juxtaposed peaks. In other words, consider the sequence $\{O(n/2^j)\}$. This sequence is a concatenation of several bitonic sequences, each of which increases first and then decreases (for details about bitonic sequences, see [61]). Each bitonic sequence that increases first and then decreases corresponds to a peak in O^j_α. We wish to isolate the peaks that are the tallest and have the widest spread, for it is in the region over which these peaks lie that the correct value of the parameter being measured is most likely to be found. Since the characteristic function of each sensor adds an area numerically equal to the sensor's width to the area under $O(x)$, a good measure of the robustness of a peak is the area under it.

At the jth resolution, consider the sequence $\{O(n/2^j)\}_n$. This is a finite sequence since the support of O is finite. Let there be p peaks (or p bitonic sequences) in O^j. Thus, the sequence $\{O(n/2^j)\}$ can be rewritten as

$$\{O(\frac{n_0}{2^j}), O(\frac{n_0+1}{2^j}), O(\frac{n_1}{2^j}), O(\frac{n_1+1}{2^j}), \ldots, O(\frac{n_{p-1}+1}{2^j}), O(\frac{n_p}{2^j})\}, \tag{5.13}$$

where the subsequence $\{O((n_{k-1}+1/)2^j), \ldots, O(n_k/2^j)\}$ is the kth bitonic sequence from the left. Therefore, the area under this peak is given by $1/2^j \sum_{n=n_{k-1}}^{n_k} O(n/2^j)$. Since the factor $1/2^j$ is common to the areas of all peaks at the jth resolution, we may make the area scale free by dropping this factor and writing the area of the kth peak at level j as

$$A^j(k) = \sum_{n=n_{k-1}}^{n_k} O(n/2^j). \tag{5.14}$$

We then select the peak with the largest area and ignore the other peaks. The function O is further resolved over the regions over which these largest peaks occur, and the process is repeated until a satisfactory region of the real line is isolated as the most likely candidate for containing the correct value of the parameter being measured by the sensors. However, before resolving a certain selected peak further, we correct the region over which the resolution is to be carried out by adding a

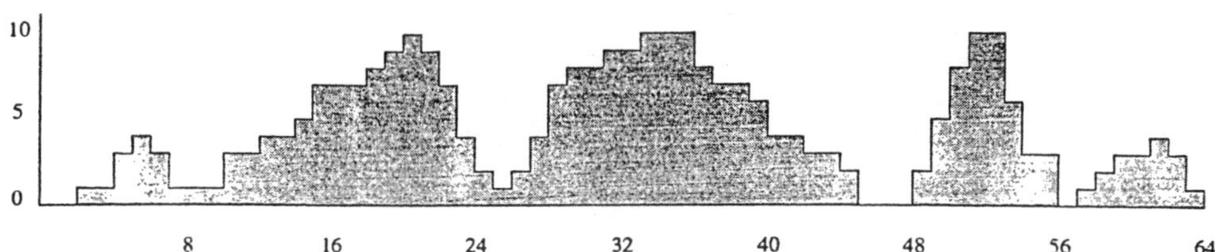

Figure 5.6 $O(x)$: shaded region indicates portion to be resolved over.

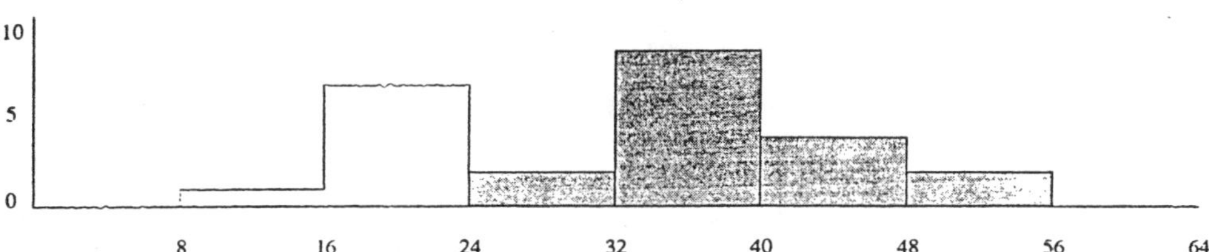

Figure 5.7 $O^{-3}(x)$: shaded region and region of width 8 at left to be resolved over.

segment of length $1/2^j$. Figures. 5.6 through 5.10 show the property of the coarse-grain to fine-grain scheme for isolating robust peaks.

If at the jth resolution the kth peak is selected as the peak with the largest area under it, then the region over which the resolution of $O(x)$ is performed again is $[(n_{k-1}+1)/2^j, n_k/2^j]$ with a correction of length $1/2^j$ at the left.

Therefore, $O(x)$ is resolved over $[n_{k-1}/2^j, n_k/2^j]$. This process is continued until the interval to be resolved is smaller than the maximum acceptable width for output estimate or when further resolution does not reduce interval width. The final corrected region with the largest value of O over it is accepted as the final output estimate.

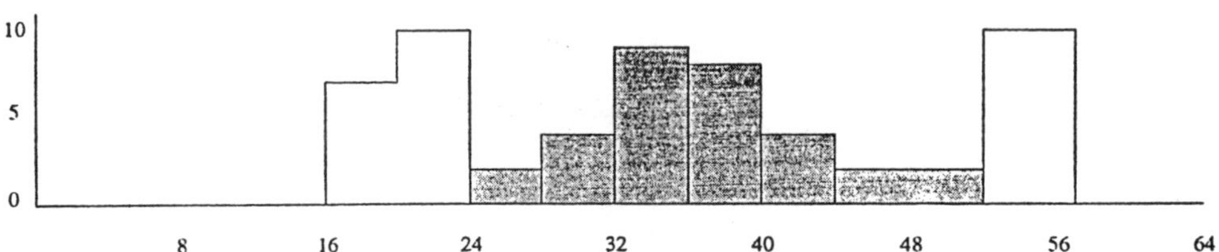

Figure 5.8 $O^{-2}(x)$: shaded region and region of width 4 at left to be resolved over.

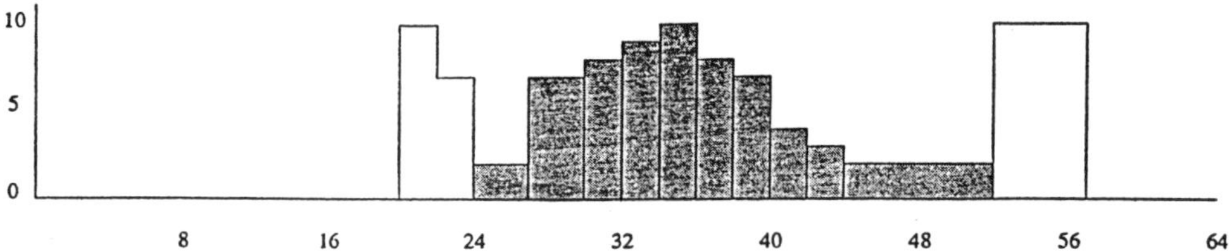

Figure 5.9 $O^{-1}(x)$: shaded region and region of width 2 at left to be resolved over.

5.2.3 Algorithm PIKR

The inputs to the PIKR algorithm (Prasad, Iyengar, Kashyap and Rao) are the end points a_i, b_i of the interval estimate $[a_i, b_i]$ of the sensors S_i, $1 \le i \le N$, and the lower and upper bounds of resolution $j_0 = \lceil \log(1/l) \rceil$ and $j_1 > \lceil \log(2N/\mathrm{Supp}(O(x))) \rceil$. The algorithm is shown below.

    Algorithm PIKR :
begin

1. Construct the array of ordered pairs:

$$[(a_1, 1), (b_1, -1), (a_2, 1), (b_2, -1), \ldots, (a_N, 1), (b_N, -1)]$$

2. Sort this array in increasing order with respect to the first component of the ordered pairs to obtain the array $[(\alpha_1, \sigma_1), (\alpha_2, \sigma_2), \ldots, (\alpha_{2N}, \sigma_{2N})]$, where

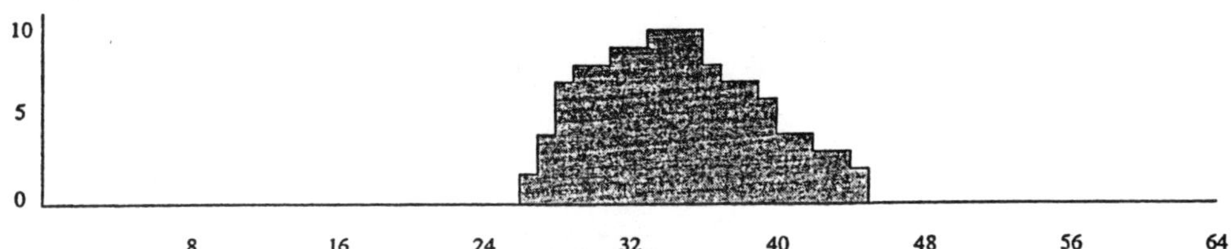

Figure 5.10 $O^0(x)$: shaded region and region of width 1 at left to be resolved over. At this point, we may terminate resolution and choose the interval over which this peak attains a maximum as the final output.

each α_i is some a_j or b_j, $\alpha_i, \leq \alpha_{i+1}$ $1 \leq i \leq 2N$, and

$$\sigma_i = \left\{ \begin{array}{ll} 1 & \text{if } \alpha_i \text{ is an } a_j \\ 0 & \text{if } \alpha_i \text{ is a } b_j \end{array} \right. .$$

3. Construct the array $[(-\infty, 0), (\alpha_1, \sigma_1), \ldots, (\alpha_{2N}, \sum_{j=1}^{2N} \sigma_j), (\infty, 0)]$ representing the overlap function $O(x)$. Note that $\sigma_1 = +1$, $\sum_{j=1}^{i} \sigma_j = 0$, and $O(x) = \sum_{j=1}^{i} \sigma_j \ \forall \ x \ \alpha_i \leq x < \alpha_{i+1} \ 0 \leq i \leq 2N$, where $\alpha_0 = -\infty$ and $\alpha_{2N+1} = \infty$. Set $n_{j_0} = \lfloor 2^{j_0} \alpha_1 \rfloor$ and $n'_{j_0} = \lceil 2^{j_0} \alpha_{2N} \rceil$.

4. **while** $j_0 \leq i < j_1$ **do** $RESOLVE[i; n_{i-1}, n'_{i-1}]$; $i \leftarrow i + 1$; **enddo**
   Sample $O(x)$ between $n_{j_1}/2^{j_1}$ and $n'_{j_1}/2^{j_1}$ to obtain $O^{j_1}(x)$, the approximation of $O(x)$ at the jth resolution. Choose the subinterval of $[n_{j_1}/2^{j_1}, n'_{j_1}/2^{j_1}]$ over which $O^{j_1}(x)$ attains a maximum (or takes values greater than a specified value) and accept this subinterval as the integrated output estimate of the N sensor estimates.

end.

Procedure **RESOLVE**, which resolves $O(x)$ to obtain an approximation of O at the jth resolution over a given interval, is shown below. It yields the indices of two points at the jth resolution, over which the largest or most prominent peak occurs, at the jth resolution.

   procedure $RESOLVE[j; n_{j-1}, n'_{j-1}]$
begin

1. Resolve $O(x)$ at scale 2^{-j} by sampling it over the interval $[(n_{j-1}-1)/2^{j-1}, n'_{j-1}/2^{j-1}]$ at the points $(2n_{j-1}-2)/2^j, (2n_{j-1}-1)/2^j, \ldots, 2n'_{j-1}/2^j$ to obtain $O^j(x)$, the approximation of $O(x)$ at the jth resolution, represented by the array

$$[(-\infty, 0), \ldots, (n/2^j, \sum_{j=1}^{k_n} \sigma_j), \ldots, (\infty, 0)], \quad (2n_{j-1}-2 \le n \le 2n'_{j-1})$$

   and $O^j(x) = \sum_{j=1}^{k_n} \sigma_j$, where $n/2^j \le x \le (n+1)/2^j$ and $(n/2^j - \alpha_{k_n}) < 1/2^j$.

2. Choose n_j and n'_j, where $n_j < n'_j$ and $2n_{j-1} - 2 \le n_j \le 2n'_{j-1}$ such that $\{O(n/2^j)\}_{n=n_j}^{n'_j}$ is a contiguous bitonic subsequence of $\{O(n/2^j)\}_{n=2n_{j-1}-2}^{2n'_{j-1}}$, which first increases and then decreases, and which has the largest sum that is, the subsequence with the maximum sum among all such bitonic subsequences.

end.

Figures. 5.6 through 5.10 indicate the various stages during the execution of this algorithm graphically.

5.2.4   Experimental Validation

The implementation of algorithm PIKR is accompanied by a parameter-driven simulator to evaluate the performance of the algorithm. Input parameters to the simulator include the number of sensors, the numbers of tamely faulty and wildly faulty sensors, the upper and lower limits for the sensor interval widths, and the correct value.

The simulator generates random intervals for the sensors. It computes the overlap function from a sorted list of the extreme points of all the sensors. The overlap function is then sampled at increasing levels of resolution, with the sampling frequency gradually increasing. The algorithm terminates when only one peak remains or when an arbitrarily fixed resolution is reached.

5.2.5   Computation and Sampling of the Overlap Function

The overlap function is described in detail in Section 5.2.1. The array

$$[(-\infty, 0), (\alpha_1, \sigma_1), \ldots, (\alpha_i, \sum_{j=1}^{i} \sigma_j), \ldots, (\alpha_{2N}, \sum_{j=1}^{2N} \sigma_j), (\infty, 0)]$$

represents the overlap function $O(x)$. In the above equation, α_i is an extreme point of some interval (either s_j or e_j for some j) such that $\alpha_i \le \alpha_{i+1}$, $1 \le i \le 2N$, and σ_i takes the value 1 if α_i is a start point of some interval and takes the value negative 1 if α_i is an end point of some interval.

The overlap function is sampled at increasing sampling frequencies as described in section 7. The simulation begins with the interval whose end points are $n_{j_0} =$

$2^{j_0}\alpha_1$ and $n'_{j_0} = 2^{j_0}\alpha_{2N}$. These end points are replaced by the end points of the peak with the largest area under it, found at the current sampling rate for resolution at the next higher level of resolution. The simulation then continues until no further refinement is possible or until a user-specified resolution is reached.

Figures. 5.11 through 5.15 show the results of the application of our algorithm to several instances of the sensor interval fusion problem. The pictures depict the following: (1) overlap function, (2) the correct physical value being estimated by the sensors, (3) the sampled overlap function, and (4) the peak of the sampled overlap function with the largest area under it. The vertical lines depicting the above move closer and closer at finer scales of resolution as seen in the figures.

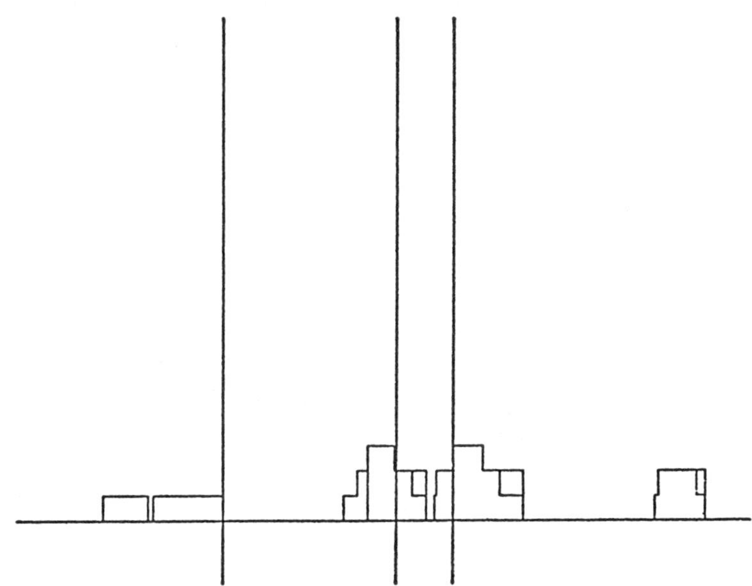

Figure 5.11 Ten sensors, 5 wild faults, 3 tame faults

We can overlay the actual sensor intervals with the overlap function.

5.2.6 Simulation Results

Tables 5.1 through 5.4 show the regions of interest in the overlap function $O(x)$ as the sampling frequency is varied from coarse to fine. Where a rise or fall of the sampled function is so sharp that it was completely missed at the current resolution level, the region of interest at the previous level of resolution is itself used at the next level. For example, in Table 5.1, the three rows marked *medium* under sampling frequency exhibit this behavior.

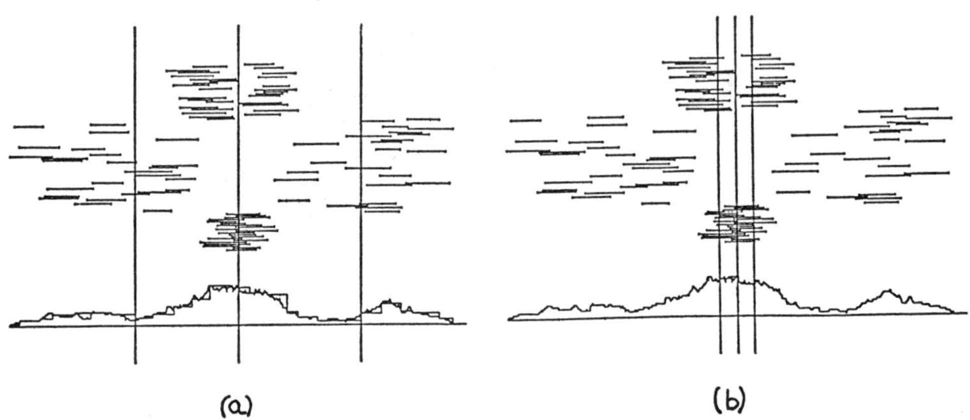

Figure 5.12 (a) 100 sensors, 50 wild faults, 30 tame faults; (b) 100 sensors, 50 wild faults, 30 tame faults.

Sampling Frequency	Resolution Level	Sensor Interval		
		Start Point	End Point	Width
Coarse	3	-0.449277	0.673841	1.123118
Medium	4	-0.191211	0.128103	0.319314
Medium	5	-0.191211	0.128103	0.319314
Medium	6	-0.191211	0.128103	0.319314
Fine	7	-0.022714	0.003010	0.025724

Table 5.1 Two Hundred sensors; 100 Wildly Faulty, 50 Tamely Faulty

Tables 5.2 and 5.3 both show the intermediate stages of resolution for 100 sensors, but with different proportions of faulty and correct sensors. It is clear from these tables that as the proportion of correct sensors increases the central peak containing the correct physical value (see Figures. 5.12 and 5.13) sustains at different levels of resolution.

Figure. 5.12 graphically shows the information summarized in Table 5.2. As seen from the figure and the table, there are only two broad scales in the overlap function, a coarse scale (first two rows of the table) and a fine scale. The correct sensor intervals all contain the correct value, the vertical green line. The tamely faulty sensor intervals are clustered at the top of the two figures located not too far

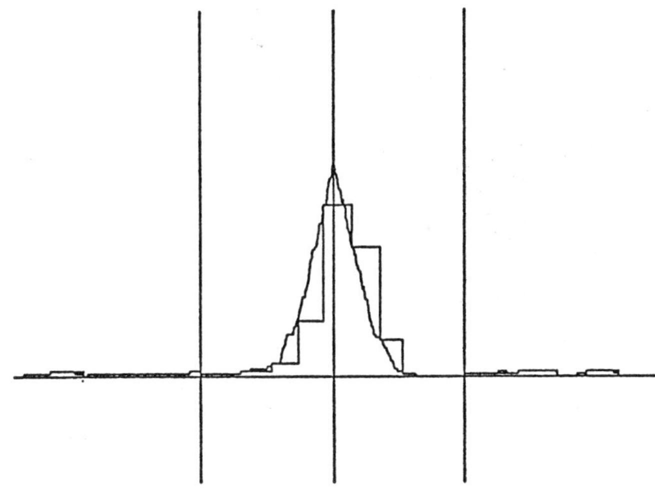

Figure 5.13 One hundred sensors, 10 wild faults, 10 tame faults

| Sampling | Resolution | Sensor Interval | | |
Frequency	Level	Start Point	End Point	Width
Coarse	3	-0.675956	0.663665	1.339621
Medium	4	-0.675956	0.663665	1.339621
Medium	5	-0.115485	0.074100	0.189585
Fine	6	-0.115485	0.074100	0.189585

Table 5.2 One Hundred Sensors: 50 Wildly Faulty, 30 Tamely Faulty

from the correct value; the wildly faulty sensor intervals are located on either side of the correct value a significant distance away from it.

As an illustration of extreme-case behavior of our algorithm, we include Figures. 5.13 and 5.14. Figure. 5.13 shows the overlap function corresponding to 100 sensors, 10 wildly faulty and 10 tamely faulty (see Table 5.3 for data). Notice that the most significant peak is sustained at every intermediate level of resolution, a phenomenon that faithfully repeats whenever the fraction of correct-valued sensors among all the sensors is relatively high.

What happens when most of the sensors are faulty is illustrated in Fig. 5.14. Table 5.2 contains the results of the simulation with 1000 sensors. Notice that the final interval computed by the algorithm does not include the correct value due to the very large fraction of wildly faulty sensors. It is relevant to mention one caveat

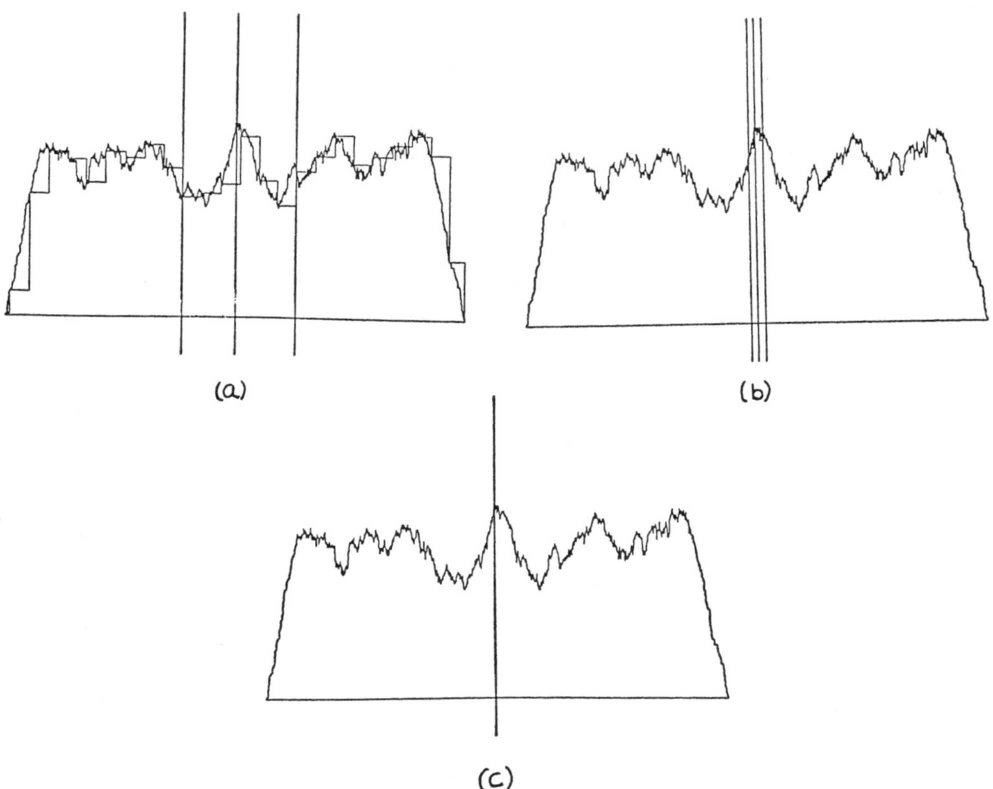

Figure **5.14** (a) 1000 sensors, 800 wild faults, 100 tame faults, (b) 1000 sensors, 800 wild faults, 100 tame faults, (c) 1000 sensors, 800 wild faults, 100 tame faults.

here. The actual form of the final solution also depend on the seed of the random number generator used to start the Monte Carlo simulation. Thus, it is not unusual for the final solution of one particular simulation experiment, with almost all the sensors being faulty, to contain the correct value.

Figure. 5.15 shows the sampled overlap function, at various stages of resolution, corresponding to the data in Table 5.1. Figure. 5.15(b) shows that the region of the overlap function sampled does not change throughout the medium frequency range, as also evidenced by the three rows in Table 5.1 marked *medium* under the sampling frequency column.

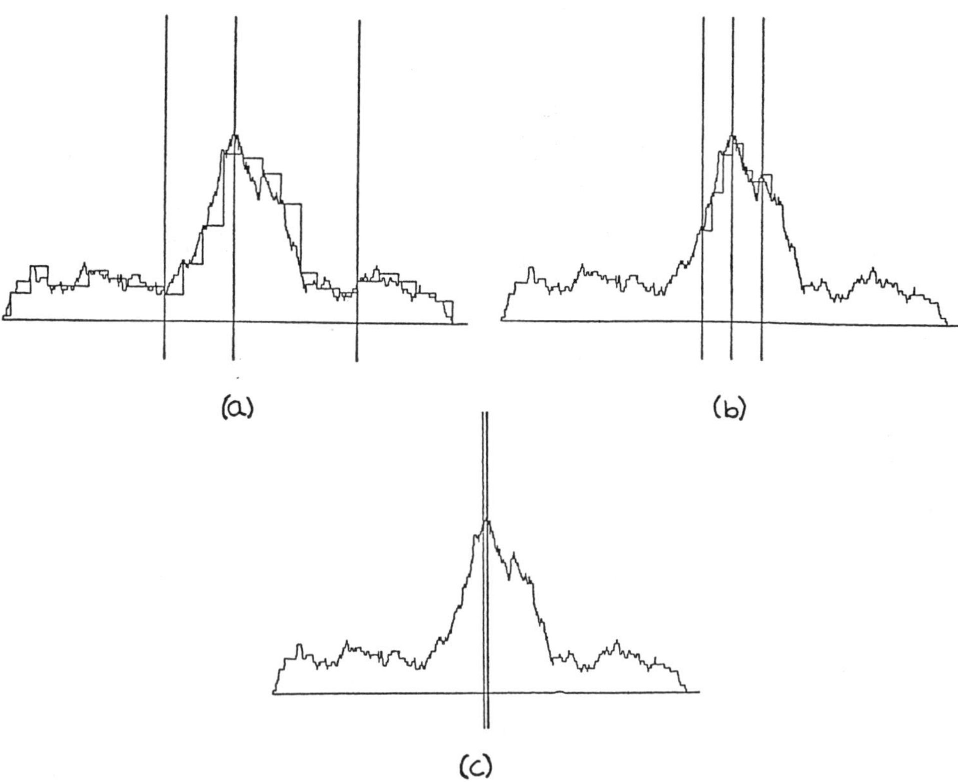

Figure 5.15 (a) 200 sensors, 100 wild faults, 50 tame faults; (b) 200 sensors, 100 wild faults, 50 tame faults; (c) 200 sensors, 100 wild faults, 50 tame faults.

5.3 Chapter Summary

Multiresolution decomposition is a powerful technique with numerous applications. We have shown its use to integrate N one-dimensional abstract sensors in a fault-tolerant manner. This is done by computing the overlap function of the sensors and resolving this function at increasingly finer dyadic scales to obtain a sequence of functions, each of which consists of a series of peaks. In each of these functions, the highest peak with the largest spread is chosen and further resolved at a finer scale to obtain the next function. This process is repeated finitely many times up to a certain scale, and the sensor estimate is taken to be the region over which the

Sampling	Resolution	Sensor Interval		
Frequency	Level	Start Point	End Point	Width
Coarse	3	-0.627078	0.379362	1.006440
Medium	4	-0.627078	0.379362	1.006440
Medium	5	-0.627078	0.379362	1.006440
Fine	6	-0.627078	0.379362	1.006440

Table 5.3 One Hundred Sensors; 10 Wildly Faulty, 10 Tamely Faulty

Sampling	Resolution	Sensor Interval		
Frequency	Level	Start Point	End Point	Width
Coarse	3	0.252377	1.006563	0.754186
Coarse	4	0.252377	1.006563	0.754186
Medium	5	-0.047790	0.203101	0.250891
Medium	6	-0.047790	0.203101	0.250891
Medium	7	-0.016398	0.021779	0.038177
Fine	8	-0.012628	-0.006013	0.006615
Fine	9	-0.012628	-0.006013	0.006615

Table 5.4 One Thousand Sensors; 800 Wildly Faulty, 100 Tamely Faulty

maximum value of the largest peak is attained.

ThePIKR method helps isolate the neighborhood of the correct value of the parameter being measured by taking advantage of the fact that the maximum intersection of intervals occurs about the correct value, and this corresponds to the highest peaks, further more, the tamely faulty sensors cluster around the correct value, contributing to the height and spread of the peak under which the correct value lies. At each resolution, only the relevant details of $O(x)$ are considered, increasing the efficiency of the computation.

5.4   References and Further Reading

The chapter is adapted from the paper [91].

Multiresolution is an important technique. For multiresolution techniques for image processing, see the seminal paper [3]. Multiresolution techniques are also used with wavelets; see, for instance, the book [93].

Problem Set 5

5-1. What is the overlap function $O(x)$ of the abstract sensors?

5-2. Show that $O(x)$ has compact support.

5-3. What is the time complexity for dynamically maintaining $O(x)$?

5-4. What is an appropriate scale for performing multiresolution on $O(x)$?

5-5. What is a bitonic sequence?

5-6. Suggest a way to construct random sensor intervals.

5-7. What is the complexity of procedure **RESOLVE**?

Chapter 6

Integration Of Redundant Multidimensional Sensors

A change in technology often cause devices to become obsolete. Sensing devices are no exception. It is important to be able to replace physical sensors with better but functionally equivalent sensors as technology progresses. The use of abstract sensors allows us to dissociate the sensor integration process from the physical devices used to sense and capture data.

In the previous chapters, we described several techniques for integrating one-dimensional abstract sensors. We spent a considerable amount of time on one-dimensional abstract sensors for two reasons.

1. It is important to explain the sensor integration process elegantly and formally. Whereever appropriate, we have presented termination proofs, proof of correctness, and bounds on the computation.

2. The one-dimensional sensor integration algorithms can be adapted for use as procedures for integrating the multidimensional abstract sensors. The one-dimensional theory can generally be extended to the multidimensional case.

6.1 Multidimensional Sensors

A fundamental problem in signal processing is to combine sensor data from redundant sensor arrays measuring multiple dimensions. For example, the position of an object is generally specified by a point, say (x, y, z) in three-dimensional space. Its velocity can be represented a three-dimensional vector, say (v_x, v_y, v_z). Suppose we want to represent both the position and velocity explicitly, say (x, y, z, v_x, v_y, v_z), at a given point. Then the data can be considered as six dimensional. [1]

As in one-dimensional case, multidimensional sensors can provide inconsistent data or even be faulty, Hence, the sensor readings cannot be blindly averaged. Efficient multidimensional sensor integration techniques are needed to resolve inconsis-

---

[1] To represent velocities of an object at all points in a three-dimensional space, one might also use a three-dimensional matrix, where each element is a vector (v_x, v_y, v_z).

tencies and to increase fault tolerance. Their performance is crucial to automated distributed tracking and detection systems.

Chew and Marzullo [13] describe several techniques for the averaging of N D-dimensional sensors. Their work is a generalization of one-dimensional sensor integration discussed by Marzullo [53].

Brooks and Iyengar [9] generalize Jayasimha's work [34] on one-dimensional sensor integration. Brooks and Iyengar [90] give an alternative technique based on K-dimensional trees. They have also developed a mathematical framework for multisensor fusion. [2]

6.2   Chew and Marzullo's Method

Marzullo [53] differentiates between a concrete sensor and an abstract sensor. His model of sensor integration assumes that of the given N one-dimensional abstract sensors no more than F can be faulty. His algorithm M1 integrates the N given abstract sensors and returns a reliable abstract estimate, which is the smallest interval containing the actual value. Recall that Marzullo finds all possible $(N - F)$-cliques in the interval graph corresponding to the N abstract sensors and returns a contiguous level that spans the least upper bound and greatest lower bound of the $(N - F)$-intersecting intervals. It is a safe estimate, but the width of the output estimate could grow large with increasing N.

Chew and Marzullo [13] consider the general problem of integrating D-dimensional abstract sensors. The case for $D = 2$ might be treated specially by algorithms that do not generalize to higher dimensions. Thus, we are interested in cases where $D > 2$.

Chew and Marzullo list four desirable properties of the abstract estimate returned by the sensor integration:

1. Assuming no more than F failures out of the given N sensors, a reliable abstract sensor estimate should be guaranteed to contain the true physical value.

2. The estimate should have a shape that is within the same class as the shapes delivered by the individual abstract sensors.

3. The estimate should be accurate.

4. The estimate should be efficient to compute.

Properties 3 and 4 are contradictory and might need trade-off. We generally require properties 1 and 2.

Chew and Marzullo define $I_{[F,N]}(S)$ to be the smallest region that satisfies properties 1 and 2. We will write I_s when F and N are implicitly known.

---

[2] A sequel to this book is under preparation by Brooks and Iyengar.

The first property says that the resulting estimate must be safe. This is fine; however, a safe estimate that is too wide can be considered as being not accurate for some real-life applications.

The second property says that the integration process should preserve the shape of the input. For example, if the given sensors look like rectangles, the output estimate should be a rectangle. Note, however,that it is difficult if not impossible to preserve the output shape. Moreover, in preserving shape we can incur an overestimate of the integrated output.

Note that algorithm M1 returns an output that satisfies properties 1 and 2. Since the given inputs are contiguous intervals, M1 returns a contiguous interval to preserve the shape. However, this results in an overestimate of the integrated output, because the integrated output includes subintervals that cannot contain the actual value.

Chew and Marzullo observe that their one-dimensional results could be used directly for multi-dimensional sensors by simply projecting the region for sensor measurement S_i onto each of the D orthogonal axes. They call their inputs D-rectangles.

Their technique uses the *divide and conquer* paradigm. The problem (for the D-dimensional case) is broken up into D subproblems, each involving one-dimensional sensor integration. The subproblems can be solved using a suitable algorithm (say algorithm M1) and then recombined to produce a D-rectangle, or the projection rectangle.

However, there are several possible disadvantages to this approach:

1. Some information may be lost. For example, we cannot correlate the coordinates and discard some measurements that we judge to be invalid.

2. A D-rectangle is not necessarily the desired shape. An application might require the abstract estimate to be a different shape, for example, a circle.

3. The size of the integrated estimate may be larger than necessary.

Chew and Marzullo [13] showed that a reliable abstract sensor may be constructed for N sensors returning D-rectangle readings of varying sizes only when the number of faulty sensors $F < N/2D$. They also proposed an algorithm that consists of reducing the problem into D 1-dimensional problems and produces a reliable abstract sensor within $O(DN \log N)$. This method has the drawback of returning a region that is frequently larger than optimal.

Chew and Marzullo showed that the projection algorithm can be used for averaging sensors specified as D-dimensional rectangles and circles; the integrated outputs preserve the shapes. D-circles are better than D-rectangles in the sense that the bound on the size of $I_{[F,N]}(S)$ for circles grows more slowly than the corresponding bound for rectangles. However, D-circles are worse than D-rectangles in the sense that $I_{[F,N]}(S)$ is more difficult to compute for circles than for rectangles. Chew and Marzullo's results are given in Table 6.1.

Table 6.1 Comparison of Techniques

Geometry	N	Complexity	Comments
Linear	$2F + 1$	$O(N \log N)$	
Rectangles	$4F + 1$	$O(N \log N)$	
D-rectangles	$2DF + 1$	unacceptable	with $I_{[F,N]}(S)$
D-rectangles	$2DF + 1$	$O(DN \log N)$	with projection
Circles	$3F + 1$	$O(N^2)$	randomized
D-circles	$(D+1)F + 1$	unacceptable	
D-rectangles	$2F + 1$	$O(DN \log N)$	uniform size

The method has an $O(N \log N)$ time complexity for the linear and rectangle geometry. It is $O(DN \log N)$ for D-rectangles using either the projection technique or having uniform size.

Chew and Marzullo's method is simple, straightforward, and relatively easy to implement. However, it has several limitations.

- There is a single output, with no associated reliability measure. Since the safest estimate is given, it is generally too wide and can be improved on.

- The method is not robust against noise.

- There is no trade-off between computation time and output quality.

6.3   Brooks and Iyengar's Method

Sensor readings are considered accurate when they return the value of the physical variable they measure within a reasonable range. To increase system reliability, it is possible to use an array of sensors returning redundant data. One way of doing this is to use a voting algorithm that accepts the value given by a plurality of the sensors in an array [73]. A more accurate method is to take the readings of all the sensors and average them. This allows a provably accurate reading within reasonable bounds to be made, even in the presence of faulty sensors within the array.

Algorithm BI, proposed by Brooks and Iyengar, averages the given redundant D-dimensional sensor data in a fault-tolerant manner. It also resolves interdimensional conflicts within feasible time bounds for more than two dimensions. It is efficient and returns readings that are closer to the optimal than existing algorithms.

The terms *region* and *interval* are used to refer to a range of values where readings from different sensors intersect. The term *clique* refers to the set of sensors whose readings intersect at a region (Fig. 6.1).

Algorithm BI considers arrays of sensors that return data in D dimensions. In

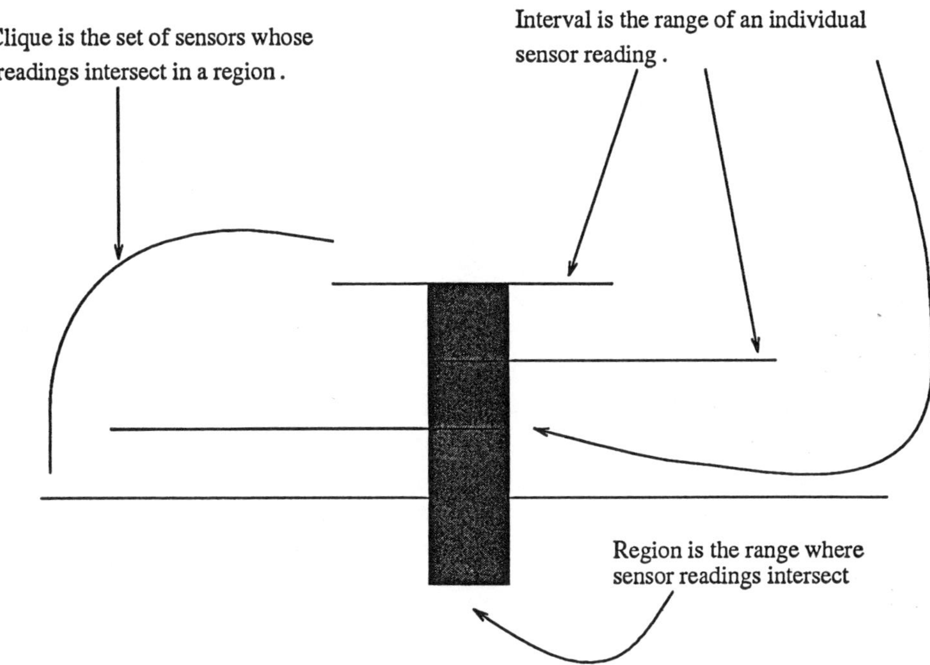

Figure 6.1 A clique is a set of sensors whose readings intersect at a region.

Chew and Marzullo's terminology, they are *D-rectangles*, that is, readings of one upper and one lower bound in each of the D orthogonal dimensions.

We assume that each sensor reading is a *triplet*:

1. Sensor identification with an associated characteristic number. For sensor j this number would be $2^{(j-1)}$.

2. A vector containing the lower bounds: $(l_1, l_2, \ldots, l_i, \ldots, l_d)$

3. A vector containing the upper bounds: $(u_1, u_2, \ldots, u_i, \ldots, u_d)$

l_i and u_i are the lower and upper bounds, respectively, of the reading for dimension i. Both vectors are of dimension D.

Algorithm J1, proposed by Jayasimha, deals with the one-dimensional in $O(N \log N)$. J1 returns a list of all regions containing cliques of at least $(N - F)$ members. It can also identify the faulty sensors in the system.

Jayasimha has shown that there are at most $2F + 1$ cliques of size $\geq (N - F)$. Marzullo has shown that the number of faulty sensors F must be less than $N/2D$

for the size of the region returned by the reliable abstract sensor to be bounded. From the above, algorithm BI assumes that the number of regions per dimension is less than $(N/D + 1)$.

6.3.1   Projection algorithm

In Fig. 6.2, we give two-dimensional sensor data from sensors 1, 2, and 3. The two dimensions are simply called horizontal and vertical. In the horizontal dimension we have three regions that intersect: $\{1, 2\}$, $\{1, 3\}$, and $\{2, 3\}$. In the vertical dimension there are only two such regions: $\{1, 2\}$ and $\{2, 3\}$. The projection algorithm returns the shaded region in Fig. 6.2, since it does not verify that sensor 1's reading does not in fact intersect sensor 3's reading. Note that the optimal region shown in Fig. 6.3 is much smaller than the one in Fig. 6.2.

The projection algorithm has a time complexity of $O(DN \log N)$. This is due to applying the one-dimensional algorithm to each of the D dimensions.

Figure. 6.3 shows the area returned by the new algorithm for the same problem. It takes interdimensional dependencies into consideration. Since the clique 1,3 does not exist in both dimensions it will not match any clique in the second dimension. Only cliques 2, 3 and 1, 2 will match in both dimensions.

6.3.2   Algorithm BI

A high-level description of Algorithm BI follows:

Algorithm :

begin

   Step 1: Use a suitable one-dimensional algorithm (say J1) to treat each dimension as a one-dimensional case for sensor integration.

   Step 2: Check to see if a D-intersection occur.

   Step 3: Report results.

end

Phase I

Phase I uses a suitable one-dimensional algorithm. For their preliminary investigation, Brooks and Iyengar [9] use a modification of algorithm J1 as a procedure for handling the one-dimensional projections.

Essentially, the procedure returns a list of all the sets of $(N - i)$ cliques for data in the dimension where $0 \le i \le F$. Associated with each clique are the following:

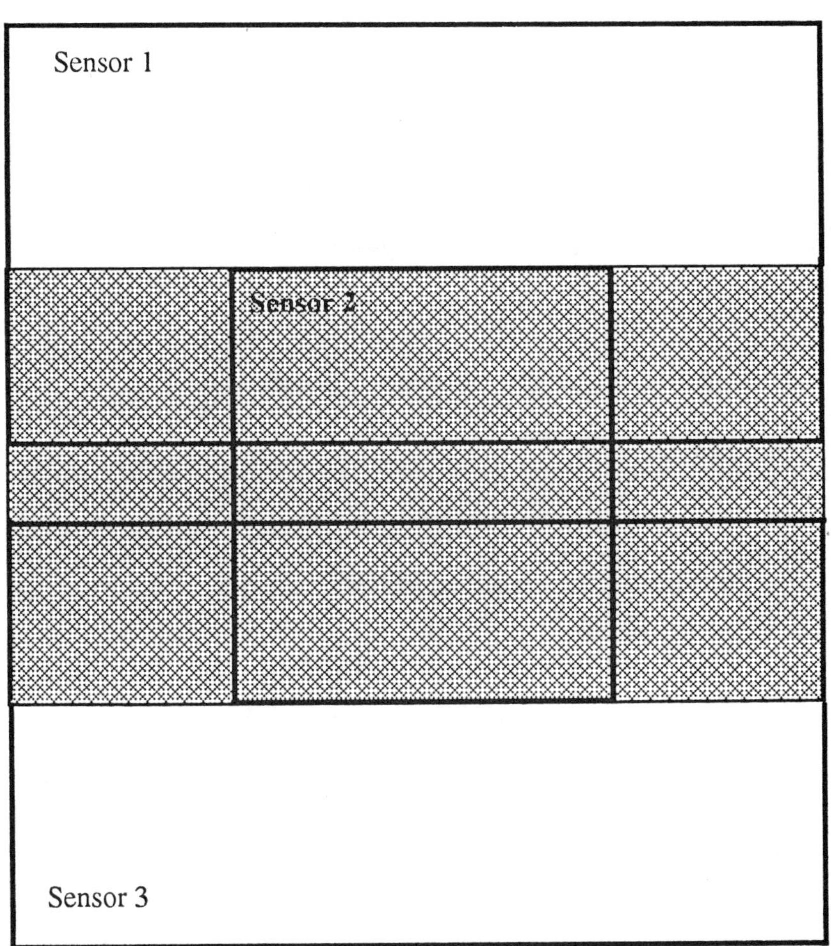

Figure 6.2 Two-dimensional problem treated by projection algorithm.

- The range (i.e., upper and lower bound) for the clique in that dimension

- A list of the members of the clique and the size of the list

- A list of the elements excluded from the clique and the size of the list

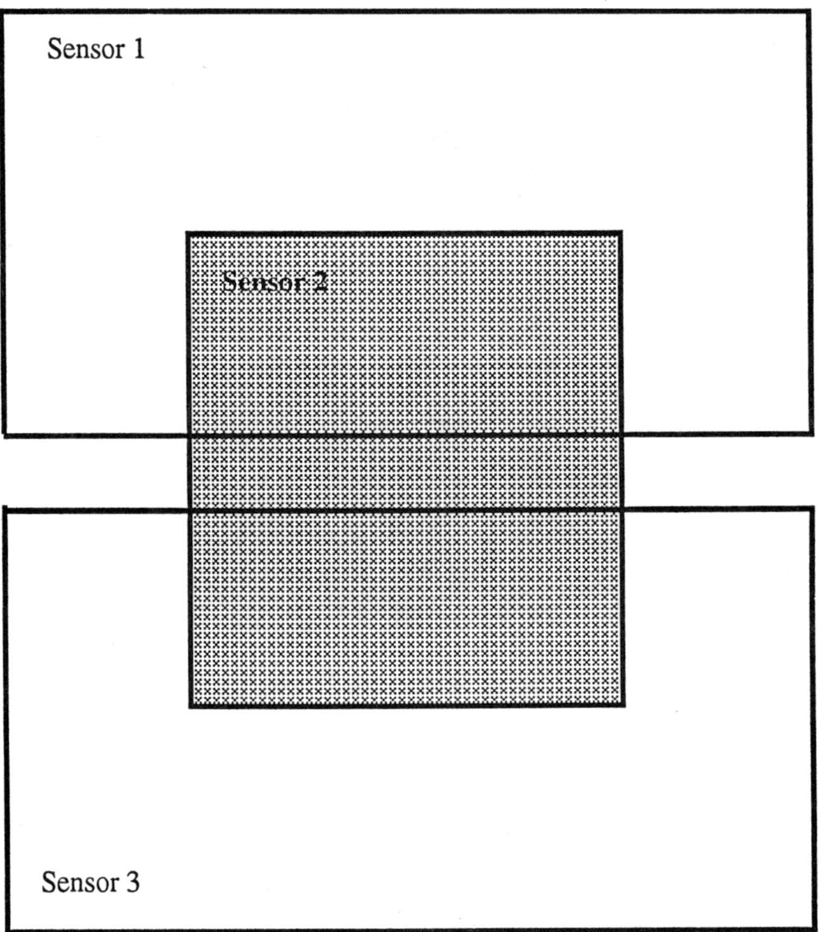

Figure 6.3 Optimal answer to the two-dimensional problem in Fig. 9.2 .

- A characteristic number for the clique, which is computed by adding the characteristic numbers of the members of the clique. (e.g., a clique of sensors 1, 4, and 7 would be represented by $2^0 + 2^3 + 2^6 = 73$)

Phase II

A D-dimensional intersection exists only when it is an intersection in all D dimensions. The cliques for each dimension are put into three lists:

- Decreasing number of clique members

- Increasing lower bounds

- Decreasing upper bounds

The cliques from each dimension are then verified against the lists for all other dimensions. The list of cliques is trimmed by removing regions that do not match any clique in at least one of the other dimensions. Matches occur when two cliques have an intersection of at least $N - F$ members.

Conditional Constraints

When any of the three following conditions are true, a match has been found and the clique is retained:

Lemma 1 *A clique of N members matches any clique of $(N-F)$ or more members.*

Proof: Since clique A contains all N elements of the universe, the elements in any other clique B are necessarily present in A. Therefore, any clique of $(N - F)$ or more elements has at least $(N - F)$ elements in common with A.

Lemma 2 *A clique of X members must match a clique of Y members if $X + Y \geq 2N - F$.*

Proof: For any two cliques A and B, they do not match if there are more than F elements that are not in the intersection. If the union of the complements of the two sets contains F or fewer elements, this is impossible. Therefore, there must exist an intersection of at least $(N - F)$ elements between the two cliques.

Lemma 3 *We designate the size of a clique, clique A, to be $N + A' - F$ and the size of another clique, clique B, to be $N + B' - F$. In the absence of the conditions stated by the two previous lemmas, A matches clique B when the number of elements in the intersection $(|XA \cap XB|)$ of the set of items excluded from A (XA) and the set of items excluded from B (XB) is greater than or equal to F minus the sum of A' and B'.*

Proof: Since neither Lemma 1 nor Lemma 2 apply, we know that the number of members in both cliques is greater than or equal to $2N - 2F$ and less than $2N - F$. A' and B' both have values between zero and $F - 1$, and the sum of A' and B' must also be less than F.

The number of elements excluded from clique A is thus $F - A'$. Similarly, $F - B'$ elements are excluded from clique B. The total number of elements excluded from the intersection of clique A and clique B must be less than or equal to F for cliques

A and B to match. If we denote the set of elements excluded from clique A as XA and the set of elements excluded from clique B as XB, we get the following formula:

$$F - A' + F - B' - |XA \cap XB| \leq F \tag{6.1}$$

which reduces to

$$F - (A' + B') \leq |XA \cap XB|. \tag{6.2}$$

Therefore, if we know the size of clique A and clique B, we can determine a match by finding the number of elements in the intersection of the sets of elements excluded from A and B.

It may be worth noting the special case of Lemma 3, which occurs when a clique (A) is of size $N - F$. In this case the clique only matches another clique when the set of sensors not included in clique B is a subset of the sensors not included in clique A. In this case, A' is zero and $F - B' = |XB|$.

Note that Lemmas 1 and 2 enable us to determine if two cliques match in constant time. If we accept operations on bit vectors as elementary operations, Lemma 3 can be implemented by doing a Boolean AND on two bit vectors and then counting the number of elements in the answer. This would then be possible in constant time as well.

If we do not accept bit vector operations as elementary operations, then Lemma 3 can be operated as an $O(F)$ operation by comparing the two sets of elements excluded from cliques A and B.

Phase III

Phase III returns the results of the algorithm. The upper and lower bounds are sorted for each dimension. The smallest lower bound and largest upper bound of the remaining cliques for each dimension define the D-rectangle returned.

6.3.3   Example

Figures.s 6.4 through 6.6 represent an example in three dimensions. Regions A and B are regions returned by one sensor that do not intersect each other. Region C is a region representing two sensor readings that intersect neither A nor B. Region D represents congruent readings from nine sensors and has intersections with A, B, and C. Notice N is 13 in this case, and since $F < N/2D$, we can tolerate at most two faulty readings.

After phase I, we have the following cliques with 11 or more ($N - F$) members and their associated ranges:

Dimension 1: {B, C, D}   12 sensors agree, from 1 to 5
                 {A, C, D}   12 sensors agree, from 9 to 13

Dimension 2: {C, D}   11 sensors agree, from 5 to 10

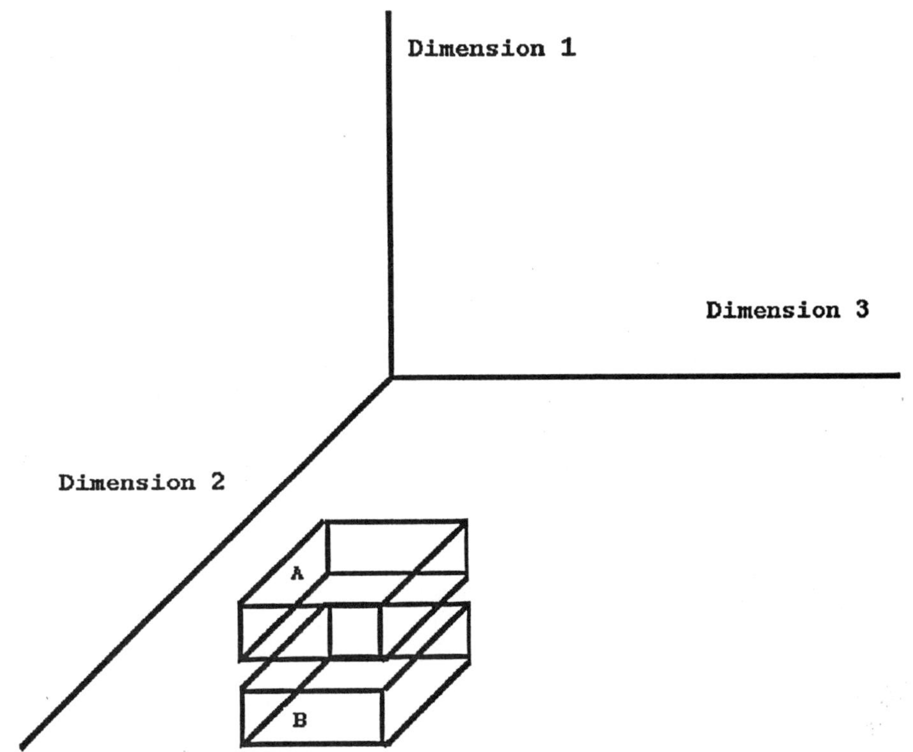

Figure 6.4 Regions *A* and *B* are disjoint and represent one sensor reading each.

$\{A, B, D\}$ 11 sensors agree, from 15 to 20

Dimension 3: $\{A, B, D\}$ 11 sensors agree, from 5 to 11
$\{A, B, C, D\}$13 sensors agree, from 8 to 11

The region that would be returned by the projection algorithm is the limits of the regions found up to now. As is shown in Fig. 6.7, this is the same as the limits of region *D* (dimension 1 from 1 to 13, dimension 2 from 5 to 20, and dimension 3 from 5 to 11).

Algorithm BI disqualifies several cliques in the next phase. In spite of this none of the cliques from dimension 1 is disqualified.

In dimension 2, clique $\{A, B, D\}$ does not have a match in dimension 1 and is discarded. Clique $\{C, D\}$ matches $\{B, C, D\}$ in dimension 1 and $\{A, B, C, D\}$ in dimension 3.

For dimension 3 $\{A, B, D\}$ does match any clique in dimension 1 and is discarded. Clique $\{A, B, C, D\}$ contains the universe and matches all cliques in all dimensions.

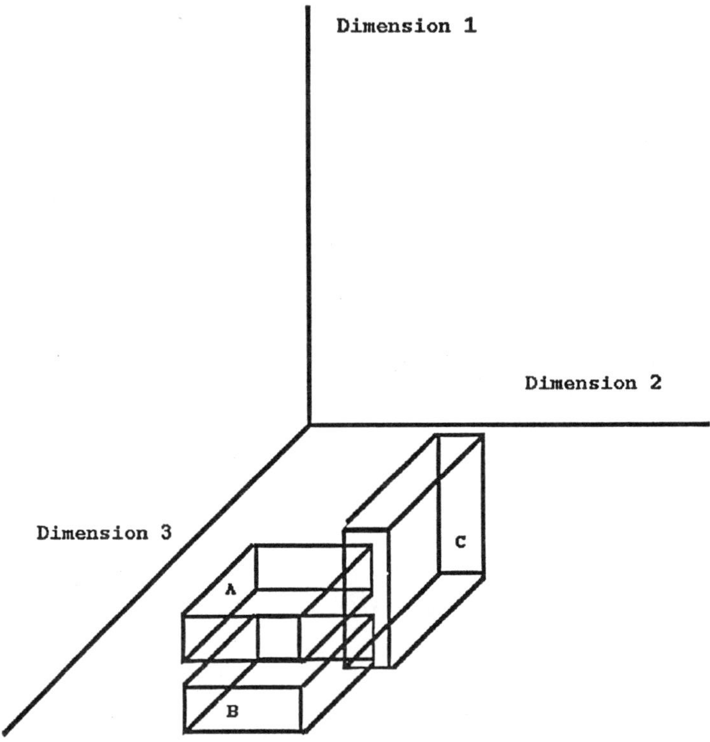

Figure 6.5 Region C represents two congruent sensor readings. A, B and C are all disjoint.

It is retained.

As is shown in Fig. 6.8, we end up with only cliques including {C, D} being retained. This is also the optimal region since {C, D} is the only three-dimensional intersection of 11 sensor readings. The region defined is dimension 1 from 1 to 13, dimension 2 from 5 to 10, and dimension 3 from 8 to 11. The set of possibly faulty sensor combinations is (B, D).

6.4 Chapter Summary

A multidimensional sensor is one that returns data in D dimensions, $D > 1$. For example, a position of an object requires three dimensions to specify it. The velocity also requires three dimensions. Thus, a sensor for position and velocity requires six dimensions.

It is important to integrate N D-dimensional sensor data. Averaging is preferred over majority voting.

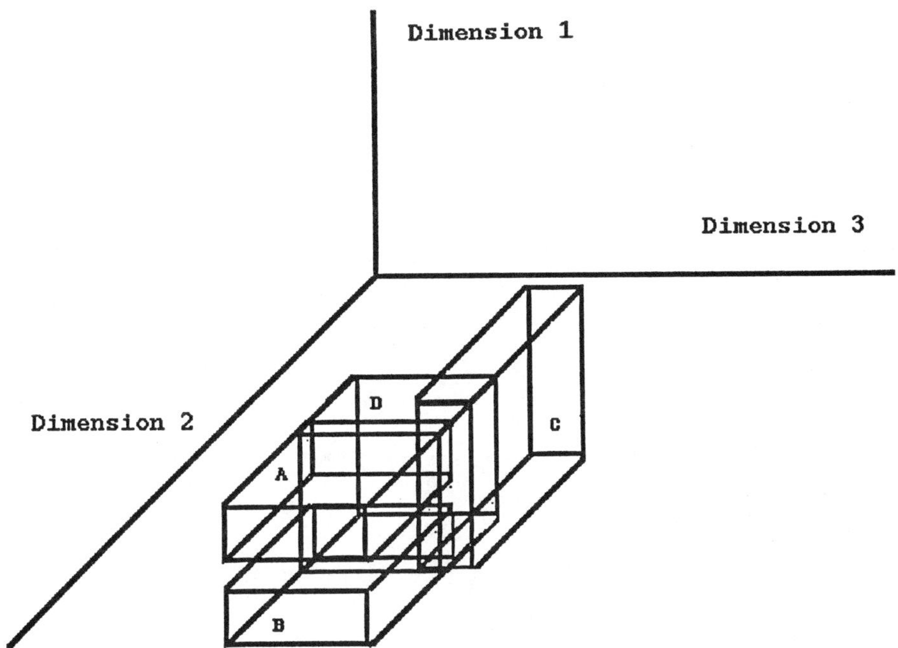

Figure 6.6 Region D represents nine congruent readings and intersects with A, B and C as shown.

Chew and Marzullo present several methods. One method deals specifically with D-rectangles and D-circles; the resulting estimate preserves the shape. The method calls for projecting the D-dimensional data onto D orthogonal axes, performing the one-dimensional sensor integration, and then combining the results to give a final estimate. Another method deals specifically with two dimensions; it cannot however be extended to higher dimensions (and still be optimal). Chew and Marzullo present four goals that should be satisfied by a reliable sensor estimate.

Brooks and Iyengar present an efficient averaging algorithm for multidimensional redundant sensor arrays providing a fault-tolerant sensor integration algorithm. It resolves interdimensional conflicts within feasible time bounds for more than two dimensions. The time complexity of the algorithm is $O(\max(DN \log N, D^2 F^2))$, where N is the number of sensors, F is the number of faulty sensors tolerated, and D is the number of dimensions.

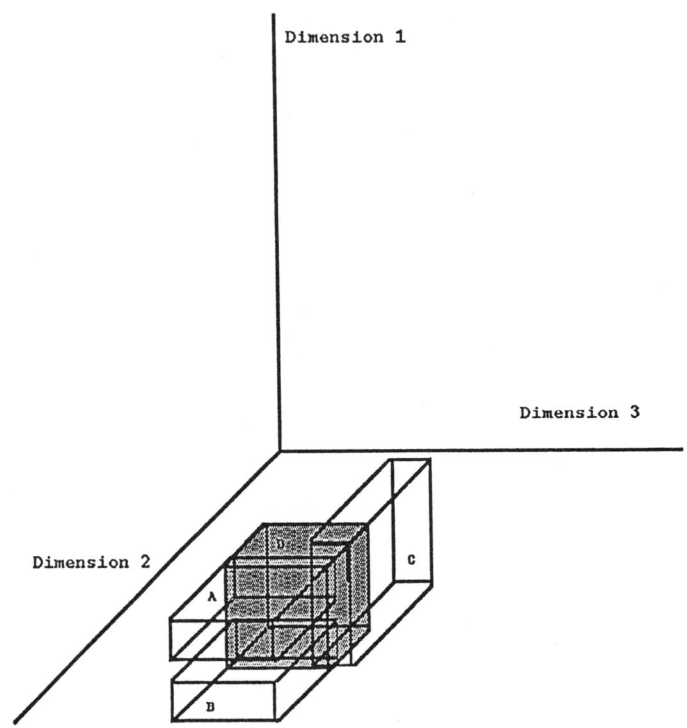

Figure 6.7 The projection algorithm returns the entire area of D.

6.5   References and Further Reading

Chew and Marzullo [13] treats the projection algorithm for the multi-dimensional case. The work is essentially an extension of Marzullo [53]

Brooks and Iyengar [9] is a generalization of Jayasimha's method [34]. Brooks and Iyengar [90] gives a theoretical framework for multidimensional sensor averaging.

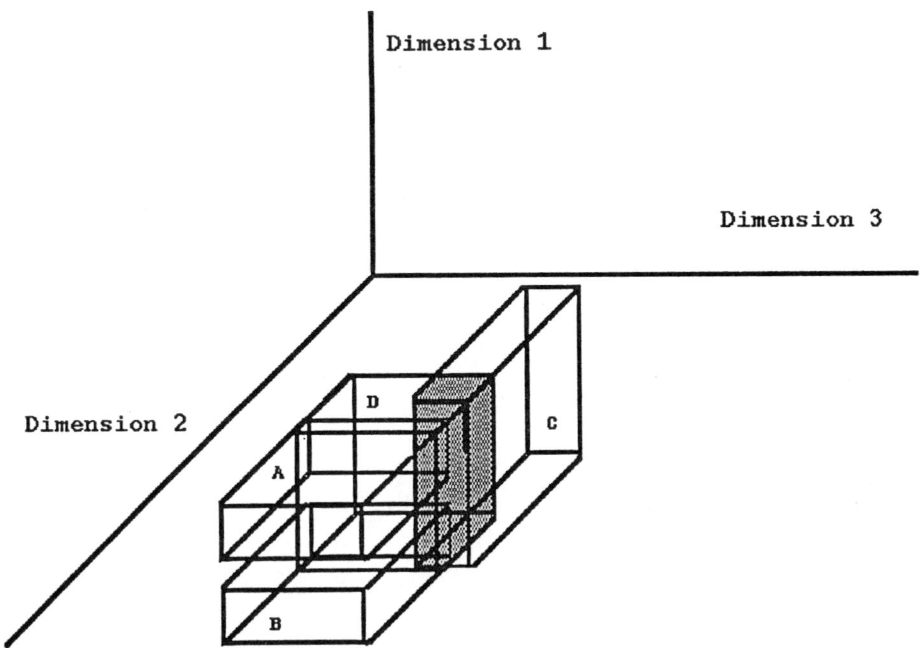

Figure 6.8 Algorithm BI returns only the intersection of C and D.

Problem Set 6

6-1. Fill in the implementation details in algorithm BI.

6-2. Show that algorithm BI is correct.

6-3. Point out one potential problem with algorithm BI.

6-4. What is the efficiency of algorithm BI?

6-5. Discuss the optimality of algorithm BI.

6-6. Compare the performance of algorithm BI with other known methods.

Table 6.2 Example system to illustrate BI algorithm.

Region Name	Number of Sensors	Dimension 1 Bounds	Dimension 2 Bounds	Dimension 3 Bounds
A	1	9 13	15 22	5 11
B	1	1 5	15 22	5 11
C	2	0 13	0 10	8 11
D	9	1 13	5 20	5 11

Chapter 7

Architectures for Sensor Integration

In the previous chapters, we described several techniques to integrate one-dimensional and D-dimensional abstract sensors. In this chapter, we will describe techniques for distributed sensor integration.

There are other interesting settings for which our algorithms do not cater directly. They cover applications for which a decision must be made on the noisy data or in a probabilistic manner (e.g., Bayesian inference). Several researchers have proposed the use of belief intervals in conjunction with the Schaffer-Dempster evidence-combining technique. Essentially, there is some measure of uncertainty and the sensor integration problem is posed as an optimization that finds the maximum-likelihood ratio in a given setting for combining the individual sensor readings.

Kadota [37] provides an excellent paper that discusses integration of complementary detection-localization systems. Two surveillance systems (with differing characteristics and grid sizes) are used to monitor an air space that can possibly be filled with thick clouds. One sensor has a higher resolution but is susceptible to atmospheric interferences; it can localize a target, but it is not sure whether it is a plane or a cloud. The other sensor has a lower resolution, but it is relatively unaffected by atmospheric interferences; it can detect a target and be sure that it is a plane (and not a cloud), but it cannot localize the target like the previous sensor. Kadota runs six algorithms and gives the simulation results. The effect of using two sensors is better than for each sensor above for the detection-localization systems.

Some would say, however, that the two radars cover the same field of interest (albeit at a different resolution or scale); hence, they provide competitive data. It is important for readers not to get carried away by jargon; the principles matter.

We will discuss algorithmic and architectural issues in distributed sensor integration. For example, we ask, What DSN topologies are appropriate for distributed sensor integration?

We have seen the widespread use of distributed multisensor data acquisition systems. This has been made possible by advances in the following two areas: signal processing, and system architecture.

Signal Processing

Numerous other researchers have worked to improve methods for handling real-time operations and for dealing with noise and interference. Significant results can be found in [6, 11, 12, 74, 75]. For a detailed treatment of advanced signal processing techniques, readers are referred to the literature.

System Architecture

Several architectures have been proposed to improve the performance and fault tolerance of complex, distributed multisensor systems. Some architectures are designed specifically to support efficient sensor integration. The focus of the chapter will be on the design and implementation of such architectures.

7.1   Architectural Issues

The design of architecture for a multisensor data acquisition system depends on several factors:

1. Network topology

2. Clock model

3. Communication

4. Routing

5. Algorithmic requirements

7.1.1   Network Topology

A graph $G = (V, E)$ is a set of vertices V and a set of edges E, where an edge connects a pair of vertices. Depending on the restrictions an edge may have, we have various classes of graphs: undirected graphs, directed graphs, directed acyclic graphs (DAGs), trees, and others. A general graph is often referred to as a *network*.

A distributed sensor network is a collection of disparate and intelligent sensors that are distributed spatially. Hence, its abstraction is a graph $G = (V, E)$, where V is a set of nodes and E is a set of edges. For a DSN, a *node* means an intelligent node consisting of a processor and associated sensors, and an *edge* refers to the connectivity of nodes.

For simplicity, we will treat a node as consisting of a single sensor/processor pair, although the techniques we discuss easily extend to the case for multiple sensors.

By network topology, we mean the way in which the sensor/processor nodes are connected. We will describe four representative topologies and discuss their advantages and limitations:

- Committee organization

- Hierarchical organization

- Flat tree

- Multilevel binary de Bruijn network (MBD)

7.1.2   Clock Model

There are two forms of *clocks*: physical clocks and logical clocks. A *physical clock* returns the actual physical time (e.g. time of the day in HH:MM:SS). Physical time has its origins in astronomy. The terms mean solar day, local sidereal time, Greenwich mean time (GMT), Gregorian calendar, and leap-year reckonings are examples of concepts of time that can be related to nature.

GMT has essentially been replaced by universal coordinated time (UCT), which uses a cesium 133 atomic clock, but also introduces leap seconds to keep in line with the astronomical time. There are several UCT servers; however, due to interferences in radio and satellite communications, it is extremely difficult to receive "precise" time.

For certain applications, the use of physical clocks is convenient but not necessary. What is important is the time order in which events happen. In such cases, logical clocks are adequate.

An example of a *logical clock* is an interval timer that counts the number of ticks (e.g., seconds or milliseconds) that have elapsed since a reference point in time. Most operating systems require the systems operator to initialize the timer with physical time; the software can easily display the timer's current reading in HH:MM:SS (or equivalent).

In the case of a timer, even though it is a logical clock, it will still return a value that is reasonably close to the physical time. However, to determine if one event happened before another event, absolute times are not needed, as explained by Lamport [45], who introduced the concept of relative times and time stamping.

Clocks are rarely perfect; they drift and become slow or fast. The drift rate should be bounded and reasonable for the clock to be useful. If there are several clocks, how can we make sure that their times agree? This involves the issue of clock synchronization, which we defer to a later section.

7.1.3   Communication

In shared-memory systems, a processor can communicate with other processors by reading to and writing from a common buffer. Even then, we have to handle mutual exclusion, deadlock, and livelock problems.

A distributed system is generally loosely coupled. Hence, a processor creates a message consisting of a message header, the sender's identification, the intended receiver(s), and the message body. It then sends the message. There are several variants of message passing depending on whether an explicit acknowledge is needed or whether acknowledgments can be piggybacked with messages.

There is computationcommunication trade-off in a distributed system. For distributed algorithms, an assumption is that the message complexity dominates other complexity measures (e.g. time complexity and space complexity). Thus, the optimality of a distributed algorithm may be expressed as a function of the message size needed to solve it.

It is interesting to note that the architectural issues may be interrelated. For instance, the choice of the topology can affect the speed and/or reliability of the message passing.

7.1.4 Routing

Given a message, how can we deliver it from the sender (or source node) to the receiver (or destination node)? This is known as the routing problem. Routing means finding a path to send the given message from the source node to the destination node.

Assume that the network is reliable and that its parameters are known and fixed. For our discussion, we have a graph with edge labels that correspond to a *cost*, for example, distance. Since the graph is static, we can use a *shortest path algorithm* to minimize the distance the message has to travel.

The distance might not be a true measure of the routing cost; the message can take a long time on a short but highly congested path. An alternative measure would be to determine which minimizes the number of hops. The key observation is that for *static routing* deterministic algorithms can be used to find an optimal path.

Now, assume that the parameters for the network can be reconfigured or are unknown. Assume further that there can be node and communication link failures in the network. Then, for dynamic routing adaptive algorithms should be used. The emphasis has shifted from speed and/or cost to reliability.

Suppose the routing algorithm encounters a failed node or communication link on the path under consideration; it must somehow find a new path (if it exists) that bypasses the failed nodes and/or communication links.

7.1.5 Algorithmic Requirements

The previous subsections can be thought of as hardware requirements for the multisensor data acquisition system. The major software requirement would be a fault-tolerant algorithm to integrate one-dimensional and multidimensional abstract sensors. We have presented several techniques to combine redundant abstract sensors. We need one major consideration before we adapt them for use in distributed sensor networks. Since there is no global clock to synchronize the sensors in a spatially distributed network, we need to check to see if the sensor readings to be integrated are "temporally close to one another"; it is not meaningful to combine sensor readings that do not correspond to the same time frame.

We will need an algorithm for competitive sensor integration in a sensor network with the following properties:

1. Ability to integrate the sensors in real time

2. Very low communication costs

3. Very low distributed computation costs

4. Fault tolerance

5. Useful, deterministic, and easy to verify and implement

The first property implies that we should use efficient data structures and algorithms so that sensor integration can be done in real time. The second and third properties both say that we want to minimize cost. Although it is desirable to keep both costs reasonably low, it might not be possible. Usually, a tradeoff between communication and computation exists. A large amount of local computation can reduce the information needed to transmit across the communication network. This is the same as saying that the bandwidth requirements in the network can be reduced by using intelligent sensors capable of local processing. Likewise, the computing load that has peaked at a node may be shared by other nodes at the expense of communication costs.

Assuming that communication costs dominate computing costs in a distributed system, algorithms should minimize the message complexity. Ideally, the message complexity should be $O(|V|)$, that is, linear in the number of nodes of the network.

The fourth property says that an algorithm should be able to withstand failures in the system. In particular, the algorithm should be fault tolerant with respect to sensor integration; it should have the ability to withstand noise and mask faulty sensors. A node failure can cause problems. A communication link failure might require a new path or route. Fault-tolerant routing assumes that the network can tolerate a prescribed amount of communication link failures.

7.2 Committee Organization

The committee organization was first discussed by Wesson et al. [80] in the context of distributed situation assessment (see Fig. 7.1).

A node consists of a processor and its associated sensors. In the committee organization, each node is autonomous. This organization is also known as anarchic organization since there appears to be anarchy, or no leadership.

Each node can broadcast information to all other nodes in the network. If all the nodes are interconnected, the committee organization is isomorphic to a completely connected graph (see Fig. 7.2).

Suppose the sensor network has N nodes, say $s_1, \ldots, s_N$. Let K_N be the completely connected graph corresponding to the N nodes. Then $\frac{N(N-1)}{2} = \Theta(N^2)$ interconnections are required. With such a high number of interconnections, this organization is robust to link failures. Messages from a source node to a destination node can easily be rerouted by bypassing the failed links.

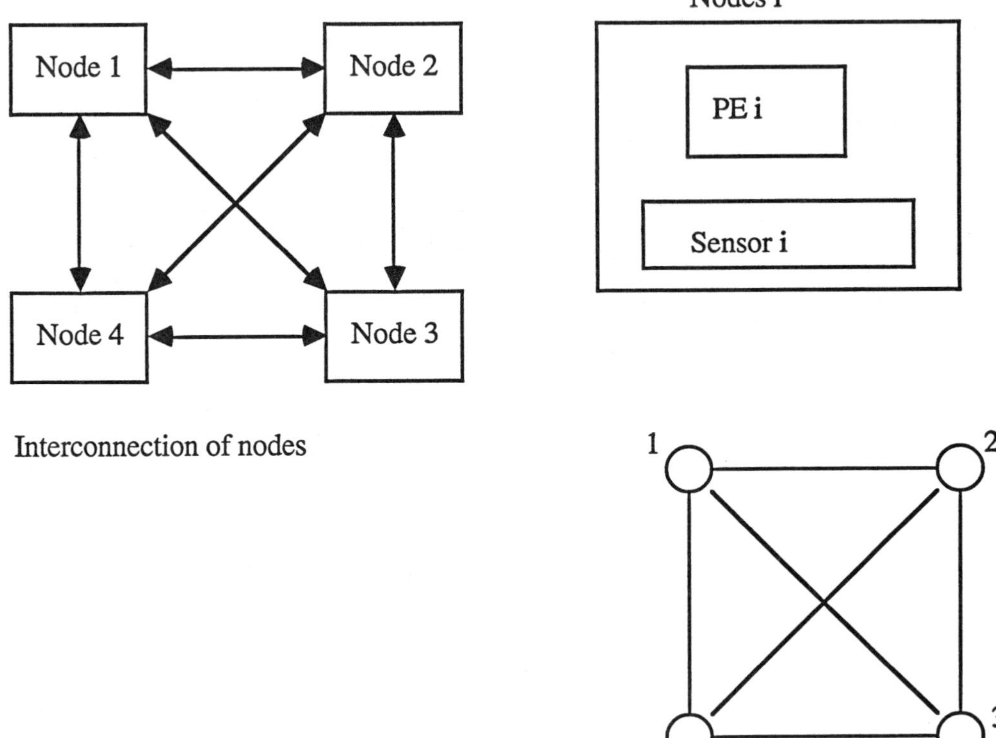

Figure 7.1 Committee organization of N nodes ($N = 4$). Each node is connected to every other node. This corresponds to a completely connected graph K_N of degree N.

For small values of N, the organization is feasible for use in a sensor network. However, for large values of N, the cost becomes exorbitant.

7.3   Hierarchical Organization

Hierarchical organization was first discussed by Wesson et al. [80] in the context of distributed situation assessment (see Fig. 7.3).

A node consists of a processor and its associated sensors. The nodes are assembled as strict hierarchies of abstraction levels. The Hierarchical organization is

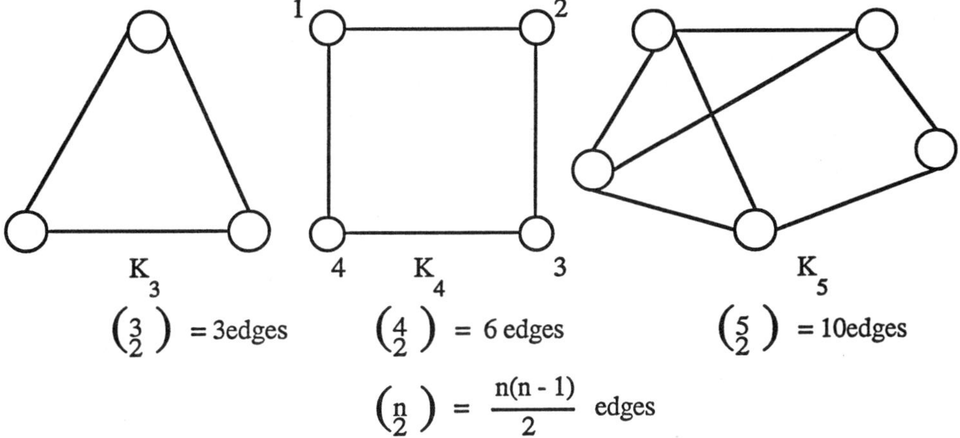

$$\binom{3}{2} = 3 \text{ edges} \qquad \binom{4}{2} = 6 \text{ edges} \qquad \binom{5}{2} = 10 \text{ edges}$$

$$\binom{n}{2} = \frac{n(n-1)}{2} \text{ edges}$$

Figure 7.2 Completely connected graphs of degrees 3, 4, and 5. They have 3, 6 and 10 edges, respectively. K_n, a completely connected graph of degree n will have $n(n-1)/2$ edges.

isomorphic to a rooted tree. There are three types of nodes: root node, intermediate node and leaf node.

At each level, nodes receive information from lower-level nodes, integrate the information received according to their position in the hierarchy, and send abstracted reports to nodes at higher levels in the hierarchy. The root or the commander is the node at the highest level. It is responsible for decision making, interpretation of data at a global level, and control of the nodes in the hierarchy.

For a sensor with N nodes, this organization requires $\Theta(N)$ interconnections. It incurs less communication costs than the committee organization. However, it could produce inaccurate estimates for two reasons. First, data sharing is not allowed between low-level sensors. Second, errors accumulate as we go up the hierarchy.

A special case of the hierarchical organization is a complete binary tree (see Figure 7.4).

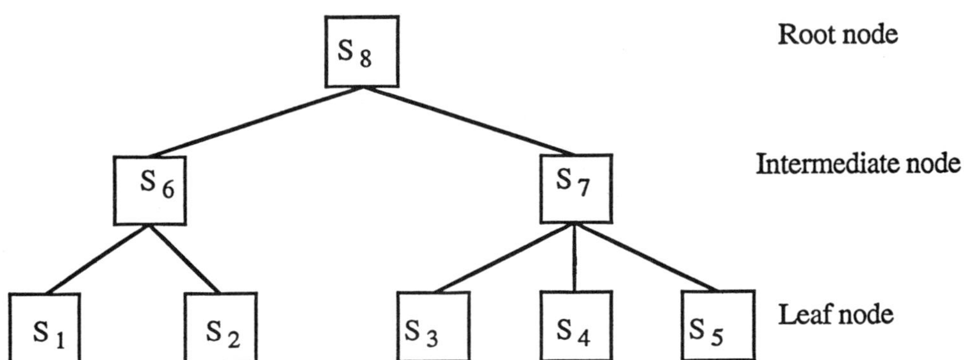

Figure 7.3 Hierarchical organization of N nodes ($N = 8$). $S_1, \ldots, S_5$ are the leaf nodes. S_6 and S_7 are the intermediate nodes. S_8 is the root node.

7.4   Flat Tree Organization

Since the committee organization and the hierarchical organization have both advantages and limitations, it is natural to ask if there are hybrid organizations that incorporate the advantages of both the committee and hierarchical organizations.

A flat tree is one possible organization (see Figure 7.5). It is isomorphic to a set of binary trees whose roots are fully interconnected [28, 31].

A *flat tree* organization has the following features.

- The sensor domain is partitioned into blocks called sensor cluster units.

- A sensor/processor node, or simply a node, consists of a processor and associated sensor(s). The sensor measures a physical variable of the environment under the direction of the processor.

- Each sensor cluster unit is internally organized as a binary tree, thus forming a hierarchy of nodes. Each cluster unit consists of a set of nodes.

- There are three types of nodes: leaf nodes, intermediate nodes, and root nodes. Note that in a pure hierarchical organization there is a single root node.

- The leaf nodes are at the first level of the network. The parent of a node is at one level higher than its children.

- Each processor in a node communicates to its children or its parent over channels through messages.

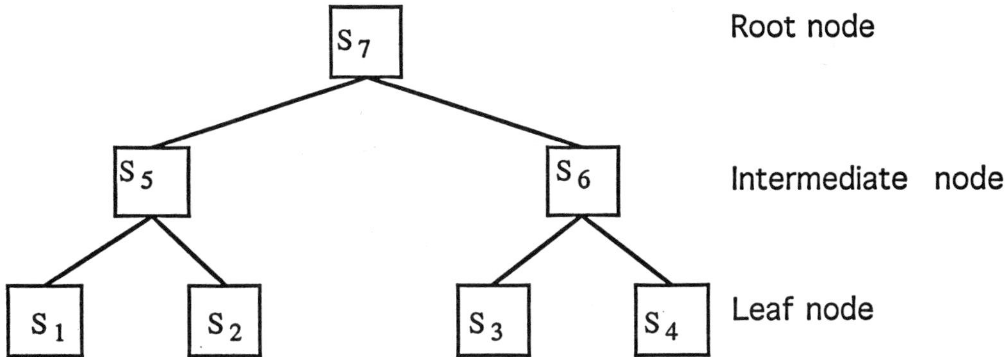

Figure 7.4 A complete binary tree with seven nodes. $S_1, \ldots, S_4$ are the leaf nodes. S_5 and S_6 are the intermediate nodes. S_7 is the root node.

- There is no global clock. A local clock runs on each processor.

- A message buffer is associated with each channel.

- The root of each tree is the commander of the respective sensor cluster unit. It acts as the control node, combining information from all other nodes in the sensor cluster unit to produce a single output.

- All commander nodes in the network are fully interconnected.

- The flat tree architecture is a hybrid of the committee organization and the hierarchical organization. The flat tree overcomes the major limitations of both the committee and the hierarchical organizations by limiting the number of communication channels required, at the same time limiting the growth of information complexity.

- The flat tree allows the information from the sensors to be integrated in real time.

- After the information is read by the sensor, the associated processor translates that information into an abstract sensor estimate, time stamps the estimate with the current time, and places the abstract estimate with the associated buffer.

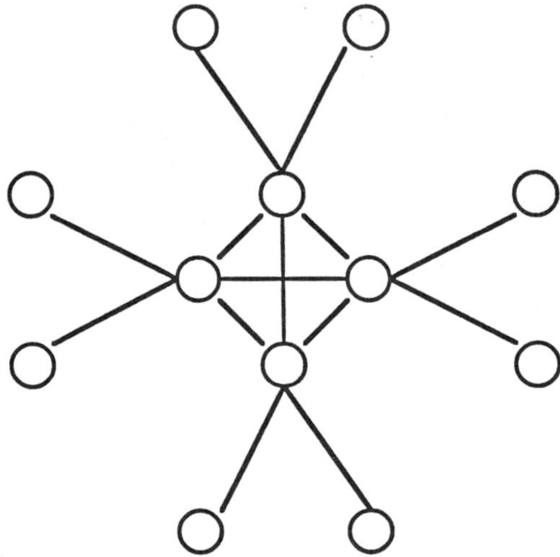

Figure 7.5 A flat tree with 12 nodes. There are four clusters. The roots of the clusters are completely interconnected. Each cluster is organized as a binary tree of three nodes.

- The flat tree is not very robust. It is possible to show that an estimate could change when a faulty sensor and a nonfaulty sensor are exchanged in one portion of the network, while such a change would not take place if these sensors are exchanged in another portion of the network [35].

- We assume that the network organization is static. In practice, however, it might be advantageous to permit dynamic reconfiguration of the network in order to offset the effects of possible node or link failures.

With the exception of root nodes, each node functions as an *abstract sensor* to produce an *abstract estimate* to send to its parent node in the sensor cluster unit tree[1]. This estimate may be either a one-interval abstract estimate (in the case of a leaf node) or a three-interval abstract estimate (in the case of an internal node).

We make two assumptions.

- Our DSN is a complete binary tree of nodes at each cluster unit and with

---

[1] The abstract estimate is essentially a range of values within which we are sure the value of the physical variable resides.

unidirectional information flow from nodes at the lower level to the nodes at the higher level. This assumption is made to minimize internodal coupling.

- The sensors are spatially distributed (with a radius of a few hundred feet), but are not geographically distant. This assumption is made to simplify clock synchronization and resynchronization issues.

The flat tree incorporates some of the merits of the committee and hierarchical organizations. nevertheless, it has two limitations.

1. Integration errors of the lower nodes accumulate as the information goes up the hierarchy.

2. It is not robust to link failures.

Performance of a Flat Tree

For the *flat tree* organization to function well, the ratio of sensor/processors in each sensor cluster unit to the total number of sensor/processors must be *small*. Because of the limited size of the sensor cluster units, the delays and message overheads are *small* and the number of levels of hierarchy is *low*.

Let N be the *total* number of sensors in the domain. Let P be an integer indicating the *average* number of sensors in a *signal processing cluster unit*. There are, on the average, N/P clusters. Then the number of interconnections needed between the commanders of each signal processing cluster unit is on the order of N^2/P^2. Thus, this method of organization achieves considerable *reduction* in bandwidth requirements.

Although increasing the number of sensor cluster units reduces the bandwidth requirements, it also increases the number of levels and time delay, as well as the number of messages within each sensor cluster unit. On the other hand, choosing a low value for P improves the performance in each sensor cluster unit, but increases the bandwidth requirements. Thus, the *choice* of P is a *trade-off* between *communication bandwidth* and *efficiency*. It is clear that the choice of P is critical.

7.5   A Versatile Architecture for Sensor Integration

We will consider a multilevel binary de Bruijn network, which will be called a MBD. It is a multilevel network with the nodes at each level interconnected as a binary de Bruijn network.

A MBD network has several good properties. It has reasonable fault tolerance. It admits simple and decentralized routing. It is easily extensible. Information integration is easily achieved even in the presence of faults.

Compared to the flat tree organization, the MBD network has better fault-tolerant properties and supports more nodes for the same diameter.

7.5.1 Operations on Addresses

We will write y^f for a binary number with bit y repeated f times. We will write $\bar{y}$ for the *complement* of y. We will use x for the *don't-care* bit. Then we can represent the binary number 00011xx by $0^3 1^2 x^2$.

A node i in a network with $N = 2$ nodes has the binary address

$$i_{k-1} i_{k-2} \ldots i_1 i_0$$

where i_{k-1} is the most significant bit and i_0 is the least significant bit. Two functions are needed to tranform addresses. Given a k-bit address, say $M = 000$, we can generate a $k+1$-bit address, say 0001, by appending a bit to it. We write app(000, 1) = 0001, where *app* is the function defined as follows: Let M be a k-bit number. Then

$$\mathrm{app}(M, y) = My.$$

Given a k-bit address, say 0010, we can generate a $k - 1$-bit address, say 001, by stripping off the least significant bit. We write str(0010, 0), where *strip* is the function defined as follows.

$$\mathrm{str}(i_{k-1} i_{k-2} \ldots i_1 i_0) = i_{k-1} i_{k-2} \ldots i_1$$

Note that $\mathrm{str}(\mathrm{app}(M, y)) = M$. After all, if we append a bit to an address and later strip it off, we should get the original address.

7.5.2 Multilevel Network

A multilevel network (MLN) is a network in which each node of the network can be associated with a level number, say m, where $0 \leq m \leq l - 1$, and l is the number of levels in the network.

Assume that a node i is at level m. Then the neighbors of node i are the set of nodes at the same level, that is, m, to which node i is connected. The parents of node i are the set of nodes at level $m-1$ to which node i is connected. The children of node i are the set of nodes at level $m+1$ to which i is connected. It is possible for some nodes to have no parents or children. For example, a root node (i.e. the node at level 0) will have no parents, and a leaf node (i.e. the node at level $l-1$) will have no children.

It is possible to have further restrictions. If each node can have at most one parent and r children, the network is called an r-ary multilevel network.

The node i at level $m(> 0)$ has the address $i_{m-1} i_{m-2} \ldots i_1 i_0$, where each digit $i_j \in \{0, 1, \ldots, r - 1\} (0 \leq j \leq m)$.

This node i is connected to at most r children nodes whose addresses are

$$\mathrm{app}(i, 0), \mathrm{app}(i, 1), \ldots, \mathrm{app}(i, r - 1)$$

and to its parent node whose address is str(i).

For every node i at level m, the relation $F_m(i)$ yields the set of nodes to which i is connected at level m. All but the 0th level of the network have the same interconnection scheme at each level.

Two nodes i and j in this network are *connected* if one of the following holds:

1. $j = \text{app}(i, b)$

2. $j = \text{str}(i)$

3. $j = F(i)$ where $b \in \{0, 1, \ldots, r - 1\}$

7.5.3 Multilevel Binary de Bruijn Network

Recall that the flat tree architecture, despite several good properties, is not robust. To overcome this problem, we can interconnect nodes in the lower levels of the network; this allows redundant communication links to tolerate failures. There are several ways to interconnect nodes. We propose the use of a de Bruijn network. This results in a multi-level binary de Bruijn network and is abbreviated MBD. A MBD is a modified one-level MLN with the top level completely connected and with each of the other levels interconnected as a de Bruijn network. The MBD network has a committee organization at each level and an overall hierarchical organization. It has several good properties:

1. It's degree is bounded (even for large networks).

2. Its diameter grows only logarithmically with the number of nodes.

3. It admits simple routing schemes.

4. It possesses fault-tolerant capabilities.

5. Its addressing complexity is low.

Using graph theoretic notation, the undirected de Bruijn network $DG(d, k)$ has $N = dk$ nodes with *diameter* k and *degree* 2^d. In binary de Bruijn networks $DG(2, k)$, which have $N = 2^k$, a node i with the binary address $a_{k-1}a_{k-2}\ldots a_1 a_0$ has neighbors (see Table 7.1):

Table 7.1 Neighbors of a Node in a MBD

LR	a_{k-2}	a_{k-3}	$\ldots$	a_1	a_0	a_{k-1}	i_1
LRC	a_{k-2}	a_{k-3}	$\ldots$	a_1	a_0	$\bar{a}_{k-1}$	i_2
RR	a_0	a_{k-1}	a_{k-2}	$\ldots$	a_2	a_1	i_3
RRC	$\bar{a}_0$	a_{k-1}	a_{k-2}	$\ldots$	a_2	a_1	i_4

The address of nodes i_1 (or i_3) is obtained by the left (or right) rotate operation i. They are called the LR and the RR neighbors of i. The address of nodes i_2 (or i_4)

is obtained by complementing the rightmost (or leftmost) bit of i_1 (or i_3). They are called the LRC and the RRC neighbors of i. Figure 7.6 shows an eight-node binary de Bruijn network. The nodes are named with the convention just described.

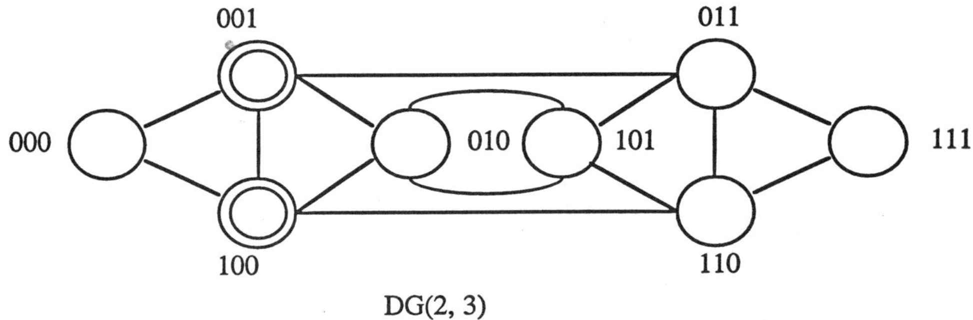

Figure 7.6 The binary de Bruijn network shown has $2^3 = 8$ nodes, diameter $= 3$, and a maximum degree $= 4$. For example, the node 001 has four neighbors 100(LR), 000(LRC), 010(RR), and 011(RRC).

An MBD network has four nodes in the topmost level. The nodes called commander nodes or root nodes are peers of each other; in other words, they are completely connected. The nodes in the underlying levels are interconnected as a binary de Bruijn network. Each node X at level m in the network is connected to two children nodes app(X, 1) and app(X, 0) at level $m + 1$ ($m < l - 1$) and is connected to its parent node str(X) at level $m - 1$ ($m > 0$).

The MBD network is a multilevel binary de Bruijn network. Since the topmost level of the MBD contains 2^2 nodes, it is convenient to assign it level 2. Hence, an l-level MBD has l levels numbered from 2 through $l + 1$.

Figure 7.7 shows a two-level MBD; the interlevel connections are shown by dashed lines and the intralevel connections by solid lines. Each node of the MBD has a PE, a clock that maintains real time, an associated sensor that samples the physical variable(s) of interest, and an associated buffer. The PE translates the sensor reading into an abstract estimate, time stamps the estimate with the current time, and places the abstract estimate in the associated buffer. There is also a buffer associated with each link. The PEs connected to the link have access to this buffer. Figure 7.8 shows the architectural details of a node of the MBD. The topological properties of the MBD are given by Lemmas 4 and 5.

Lemma 4 *The number of nodes in MBD with L levels is* $4(2^L - 1)$.

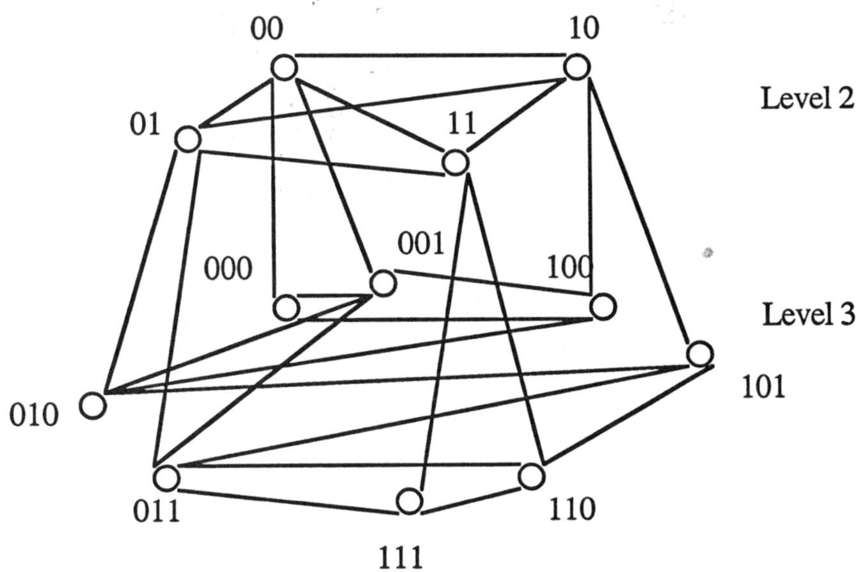

Figure 7.7 MBD with two levels called level 2 and level 3. The top level (level 2) has four nodes. The next level (level 3) has eight nodes.

Proof: The number of nodes at level m $(2 < m \leq L + 1)$ is

$$n_m = 2 \times n_{m-1}; n_2 = 4. \tag{7.1}$$

Solving equation 7.1 yields the total number of nodes as $N = 4(2^L - 1)$.

Lemma 5 *The MBD with L levels has degree 7 and diameter $L + 1$.*

Proof: The nodes at the top and bottom levels have degree of at most 5. Now consider an internal node in the network. This node is in a de Bruijn network and hence has at most four neighbors. The same node is also connected to its two children nodes and a parent node. Hence a node in the MBD has degree equal to at most 7.

For deriving the diameter of the network, consider the lowermost level in the MBD. This corresponds to DG(2, L+1) with diameter $L+1$. Note that the farthest distance between nodes in the uppermost and lowermost level is only L. Hence the farthest nodes in the MBD lie in the lowermost level; that is, the diameter equals $L + 1$. From equation 7.1, the diameter of the MBD is $O(\log N)$.

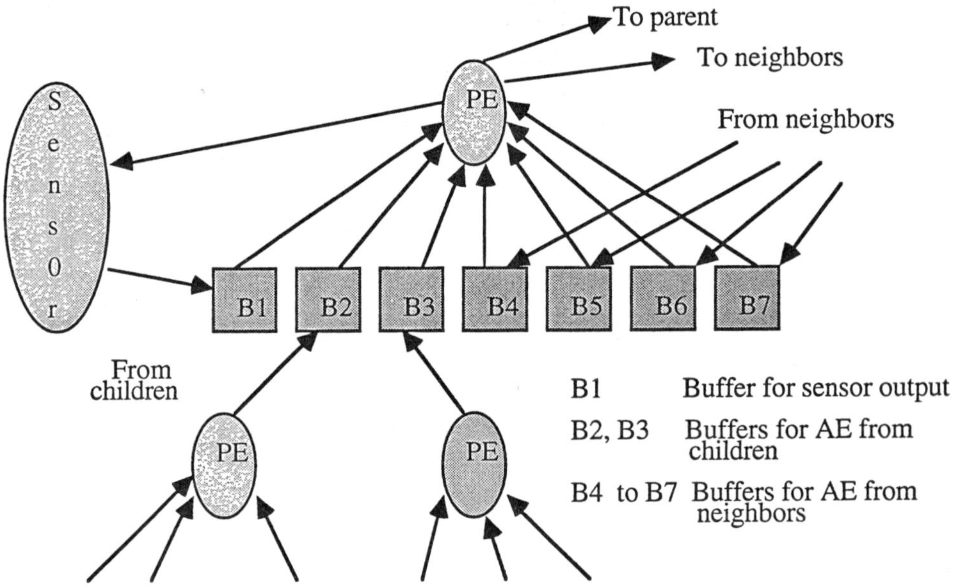

Figure 7.8 Architecture of a MBD.

7.6   Clock Synchronization

Since the physical sensor outputs typically change as a function of time, it is necessary for each estimate that is integrated to be temporally close to the others in order for the integration process to yield meaningful results. This is achieved by time stamping each estimate. If the estimates from different sensors are to be synchronized to obtain a final estimate at each commander node, then the sensors must be *sampled* at approximately the same time.

7.6.1   Clock Model

In a distributed environment, there is no *central* synchronized clock that regulates the activities of each node. Instead, each node is under the control of its *local* clock. Since a sensor responds to *real-time* activities, it is convenient for the clock at each node to provide the real time. Further more, since the estimates from different sensors have to be integrated, it is convenient to have the times provided by the sensor nodes to be *close to each other*.

The clock at each node may not be accurate for a variety of reasons, such as clock shift or change in temperature. Each clock therefore has to periodically synchronized with a more accurate clock. We assume the existence of a *central time server* on one PE of the network that, when requested for the time at t, provides the time $C(t)$.

Since the times provided by two time servers may not agree, clock synchronization with multiple time servers is more complicated than with a single time server (for a detailed treatment, see [45]).

The *central* time server itself periodically synchronizes with a *universal* time server. The latter is always accurate and lies outside our environment.

7.6.2   Clock Behavior and Synchronization Issues

Let $C_p(t)$ be the time provided by the clock on PE p at time t. Note that t itself is not observable by PE p or any of the other processing elements. Let us assume that the time $C_p(t_1)$ provided by PE p is greater than $C(t_1)$.

Let p now synchronize with the central time server. It is now possible that the new time $C_p(t_2)$ is less than $C_p(t_1)$. An abstract estimate sent out later at $C_p(t_2)$ seems as though it was sent earlier. If the abstract estimates are not integrated in a first-in first-out fashion at a node, this could lead to problems. Hence, we require that the time on a PE always *increase monotonically*. This monotonicity could be achieved by speeding up or slowing down a PE's clock each time a correction is required on synchronizing with the central time server [25]. From the foregoing discussion, we can now state the following requirements for proper synchronization (the subscript p refers to the PE p):

Requirement 1: Correct Time: The deviation in time of each clock is bounded by

$$|t - C_p(t)| \leq \epsilon_p << 1 \qquad (7.2)$$
$$|t - C(t)| \leq \alpha << 1 \qquad (7.3)$$

where ϵ_p is the maximum allowable deviation in time of a clock on a PE, and α is the maximum allowable deviation in time of the clock on a central time server.

Requirement 2: Correct Rate: Between resynchronizations, the drift rate of the clock is bounded:

$$\left| \frac{dC_p(t)}{dt} - 1 \right| \leq \kappa_p << 1 \qquad (7.4)$$

$$\left| \frac{dC(t)}{dt} - 1 \right| \leq \sigma << 1 \qquad (7.5)$$

where κ_p is the maximum allowable drift rate in time of a clock on a PE, and σ is the maximum allowable drift rate of the clock on the central time server. We assume that the clocks run continuously rather than in discrete "ticks." Hence, $dC(t)/dt$ denotes the rate at which the clock is running at time t.

Requirement 3: The clock on each PE's and the central time server increase monotonically.

We have assumed that the quantities $\epsilon_p, \kappa_p, \alpha, and \sigma$ are all fixed and known. (These quantities could be obtained from the specifications of the manufacturer's handbook.) If these quantities are time varying, then the analysis becomes very complicated.

For simplicity we assume that the constant ϵ_p is the same for all PEs and equals ϵ. Similarly, it is assumed that $\kappa_p = \kappa$. From this simplification and requirement 1, we have the following inequality. Let $C_q(t)$ be the time provided by the clock on PE q at time t.

Requirement 4: Synchronization Bound:

$$|C_p(t) - C_q(t)| \le 2\epsilon. \tag{7.6}$$

Let $\xi_{\min}$ and $\xi_{\max}$ be the minimum and maximum values of the delay in receiving the message sent by the central time server to any PE [the message contains the time $C(t)$ at time t]. Let $\delta_{\min}$ and $\delta_{\max}$ be the corresponding values for a message sent by a PE to its neighbor. Let γ be the maximum tolerance in time that a node can tolerate between intervals that can be integrated. This value of γ has to be derived from the sensor characteristics and the longest path between the leaf nodes and the commander node ($\lceil \log n \rceil$ in our case where n is the total number of PEs).

7.6.3   Clock Resynchronization

Since the clocks on the central time server and each PE drift, they have to be periodically reset. We derive a bound on the time period between resynchronizations.

Let T_s be the time period between resynchronization of the central time server. Let T_c be the time period between resynchronization of the clock on each of the PE p. The central time server resynchronizes itself every T_s seconds with a perfect universal time server which exists outside the environment of our DSN. The central time server also resynchronizes the clock on a PE every T_c seconds. Let T_s^i and T_c^i be the periods corresponding to T_s and T_c as observed by the central time server.

Lemma 6 *The period, as observed by the central time server, between synchronizations of the central time server is bounded by*

$$T_s^i \le \alpha \left(\frac{1}{\sigma} - 1 \right).$$

Proof: From the equations in the previous subsection, we can derive

$$\sigma T_s \le \alpha.$$

T_s is in the interval

$$\left[\frac{T_s^i}{1+\sigma}, \frac{T_s^i}{1-\sigma}\right].$$

Hence

$$\sigma\left(\frac{T_s^i}{1-\sigma}\right) \leq \alpha.$$

That is,

$$T_s^i \leq \alpha\left(\frac{1}{\sigma} - 1\right).$$

Observation: When a PE p is resynchronized by the central time server, at time t, there is a transmission delay of at least $\xi_{\min}$ before the value $C(t)$ reaches p. Hence

$$C_p(t') \geq C(t) + \xi_{\min}.$$

Because of the drift in the central time server, the variable transmission delay $C_p(t')$ lies in the interval

$$[C_p(t) - \sigma T_s + \xi_{\min}, C_p(t) + \sigma T_s + \xi_{\max}].$$

Hence, there could be an error of

$$(2\sigma T_s + \xi_{\max} - \xi_{\min})$$

at the time that p's clock is resynchronized.

From (7.2) and (7.3),

$$2\sigma T_s + (\xi_{\max} - \xi_{\min}) + \kappa T_c \leq \epsilon. \tag{7.7}$$

A restatement of equation 7.7 is the following theorem.

Theorem 3 *The time period between resynchronizations of the clock on a PE is bounded by*

$$T_c \leq \frac{\epsilon - 2\sigma T_s - (\xi_{\max} - \xi_{\min})}{\kappa}. \tag{7.8}$$

Proof: See the arguments in the previous paragraph.
Observation: A restatement of equation 7.8 with observable times on the central time server is (using equation 7.3 and the result of Lemma 8)

$$T_c^i \leq \frac{\epsilon - \frac{2\sigma T_s^i}{1-\sigma} - (\xi_{\max} - \xi_{\min})}{\kappa} - \alpha. \tag{7.9}$$

Using the clock and network parameters, we can rewrite equation 7.9 as follows:

$$T_c^i \leq \frac{\epsilon - 2\alpha - (\xi_{\max} - \xi_{\min})}{\kappa} - \alpha. \tag{7.10}$$

7.7   Routing

Routing involves sending a *message* from a *source* node to a *destination* node in a *network*. Routing algorithms or strategies are essential elements in the correct operation of a computer network. Several questions come to mind.

- What are the important communication constraints?

- How can we characterize the constraints?

- What is their impact on information routing?

- What are the effects of link and or node failures in a network?

- What is a suitable network structure design to overcome or minimize the constraints?

- How can dynamic routing be achieved?

A key issue in sensor networks is the *communication* between processors [81]. The principal component of a DSN is coordinated computation among the processors. Thus, communication among processors is the backbone of a DSN. Each sensor acts as a *knowledge source* and communicates to some or all other nodes in the network to initiate the inference process. The interaction among knowledge sources is an expensive operation in distributed systems. It is important to minimize interprocess communication and design efficient ways to route the information in the network.

7.7.1   Information Routing Issues in a DSN

A processing node receives a bulk of data from the associated sensor at regular intervals and generally at a fixed rate. The node processes the data and sends the information to some or all other nodes in the network, depending on the problem solving technique. Since the data generation is repetitive, it is imperative to route information in an efficient manner.

The requirements for information routing in a DSN are as follows:

- The entire information generated by a sensor should fit in a packet; otherwise, loss of a packet or delay in receiving it might lead to discarding the entire data. This is contrary to the practice used in conventional data communication networks where information is transmitted in several packets. The idea behind our requirement is to speed up the inference process and to reduce queue sizes.

- In most DSN applications, the sensor data are generated and transmitted in each sensing cycle. Since the data exchange is almost continuous, the communication protocols should be designed such that an explicit *acknowledge* is not used for each packet. This saves enormous traffic on the network considering the size of the DSN.

- By not using acknowledge messages, there can be old packets that have not yet been processed when a new packet arrives. We can simply ignore the old packet which could be identified by using time stamps in the message protocols. However, we should see that data are not lost by ignoring old packets. It is necessary to route the packets within a maximum allowable time.

- Since a DSN is envisaged to operate under hostile environments, it is therefore necessary to employ reliable point-to-point communication protocols. This topic has been well studied in the context of computer networks. These should be adapted to the DSN domains.

Communication among the cooperating processors is the backbone of a distributed sensor network. Thus, we will address the issues relating to the interprocess communication.

Each processing node faces two questions:

What to communicate? This question is complex, and problem dependent and has been discussed in [47, 74].

How to communicate? This is our main focus. There are two important communication constraints: the delay constraint and the reliability constraint.

7.7.2   Routing Techniques

There may be one or more paths from a *source* node to a *destination* node. The path with the *minimum* length is called a *shortest* path. Several *shortest*-path algorithms are known.

Optimal message routing involves considering the *time delays* in *message passing* in DSNs. In an ideal situation, where all parameters of a network are assumed to be known and unchanging, it is possible to determine a routing strategy that optimizes network performance (e.g., minimizing the average network delay). Routing strategies must also be capable of adapting to changing topologies if *task decomposition* calls for network reconfiguration.

Considering changing situations in sensor networks (such as link failure, a change in traffic distribution, or node failures), it is necessary to have some degree of adaptability in the routing algorithms. Because DSNs suffer from *volatility*, it is imperative to develop algorithms that minimize the effects of overall topological changes in the network that might result from this volatility.

Distributed Routing Algorithms

Under the problem solving strategies for a DSN, it is imperative that efficient routing schemes be developed for information dissemination in the network. Each sensor generates a bulk of data that needs to be transmitted to some or all other nodes in a DSN, depending on the problem solving technique. It is clear that the routing scheme depends on both topology and problem solving strategies. Topology changes could occur in a DSN due to link failures, interceptions, jamming, an the like.

Some fail-safe, dynamic, distributed routing protocols can be found in [4, 14, 32, 41, 54, 64]. They can be modified to suit the DSN domains.

There are basically two methods for information routing in any network.

- Providing every node with the entire topological information.

- Use routing tables containing the shortest path information.

Subtle issues in algorithms for these two methods have been discussed in [32].

Distributed topology learning algorithms have been discussed in [35, 64, 70]. Sharma et al. [30] present an efficient distributed topology learning algorithm that could be employed for a DSN. However, due to the large size of DSN, it would be prohibitive to store the topology information at every node. Also, this scheme would increase the traffic on the network whenever topology updates have to be made. (Topology update messages are sent to nodes in the networks when node or link failure occurs and when a link or node comes up.)

The schemes that use shortest paths are better suited for a DSN. However, dynamic determination of optimal routes is almost infeasible for large networks like a DSN. The hierarchical routing schemes of Kleinrock and Karmovm [41] and Baratz and Jaffe [4] focus on solving the dynamic routing problem in large networks. The main idea is as follows. The network is partitioned into set of disjoint clusters and into k levels. Shortest-path information with respect to a cluster is maintained at each node in the cluster. A node also stores an additional fact, which is the shortest-path information to each supercluster. This scheme is shown to achieve optimality in path length, asymptotically.

The hybrid topology configuration is similar to the hierarchical structure discussed above. Each sensor cluster unit in the hybrid structure could be considered as the cluster of nodes in the scheme described in [41]. Due to the small size of an SCU, we could employ the cluster shortest-path information at each node in cluster. Employing the algorithms in [4, 41], we could find an optimal routing path in a DSN. Details of these algorithms can be found in [4, 32, 41, 54].

7.7.3 Addressing Scheme in a MBD

Consider a MBD with L levels. By convention, the levels are numbered from 2 to $(L + 1)$. The address of a node in this network consists of two parts :

Level number of the MBD in which it is present. This requires $\lceil \log L \rceil$ bits for its representation.

Index of the node in that level. This requires at most $(L + 1)$ bits to index a node in any level, because the lowermost level [i.e., level $(L + 1)$) contains $2^{(L+1]}$ nodes. The address of a node in a MBD with L levels hence needs $\lceil \log L \rceil + (L + 1)$ bits.

To extend a MBD with L levels, we can add the additional nodes at the lower-most level. Thus, extending the network requires a fixed number of interconnections between the new nodes and the nodes at level $(L+1)$ only. Note that the information integration process will not be affected at any other level of the MBD. Additional bits may be needed to address the nodes in the new level.

Messages can be routed efficiently in a decentralized manner in the MBD. We first consider routing within a level (also known as intralevel routing) and then consider routing across levels (interlevel routing). To evaluate the routing complexity, we assume that a message takes unit time to traverse a link.

7.7.4 Routing in a MBD

Routing in the top level takes a unit time step since the nodes are completely connected. Routing in a de Bruijn network is a well studied problem [18, 59]; consider the routing algorithm presented in [59]. In this algorithm, tag bits are appended to the message at the source before routing. These tag bits are used by intermediate nodes to compute the address of the next node in the path. This method assumes that all the nodes in the path are fault free. Hence the algorithm will fail if any of the intermediate nodes or links are faulty.

Two distributed routing algorithms, PATH.1 and PATH.2, are described. The address of the next node is computed at the previous node in the path. PATH.1 takes $\Theta(\log N)$ steps in a de Bruijn network with N nodes, and PATH.2 takes $O(\log N)$ steps.

Let a binary de Bruijn network have $N = 2^k$ nodes and let $S = s_{k-1}s_{k-2}\ldots s_1 s_0$ be the *source* node that sends a message to the destination node $D = d_{k-1}d_{k-2}\ldots d_1 d_0$. The *message* consists of the *data* and the *message header*.

The *message header* contains a *routing tag* whose content depends on the type of routing being performed. Two types of routing tags are used:

Type 1 tag	Source node address	Destination node address	Counter 2	Interlevel routing bit, IRB

Figure 7.9 Type 1 routing tag consists of type, source address, destination address, routing tag counter, and IRB.

Type 1: This signifies *normal* routing. The Type 1 routing tag contains the following:

1. Source node address
2. Destination node address
3. Counter z used for
   (a) storing the number of message hops from the source node to the current node
   (b) generating the address of the next node in the path
4. Interlevel routing bit (IRB), which is
   (a) set if the source and the destination nodes are in different levels, or
   (b) reset if the source and destination nodes are in the same level.

Figure 7.9 shows a Type 1 routing tag.

Type 2: This signifies *fault tolerant* routing. When the routing algorithm encounters a fault, rerouting information is appended to a type 1 tag. The appended type 1 tag is called a type 2 tag.

PATH.1 Algorithm

From the construction of the de Bruijn network it follows that the source node has the following neighbors: $d_0 s_{k-1} s_{k-2} \ldots s_1$ and $s_{k-2} \ldots s_1 s_0 d_{k-1}$. Using this property, we can now generate two routes by appending successive bits of the destination node to the source address. Clearly, routes 1 and 2 take exactly $k = \log N$ steps.

Table 7.2 Route 1

$z = 0$	s_{k-1}	s_{k-2}	$\ldots$	s_1	s_0	(source)
$z = 1$	d_0	s_{k-1}	s_{k-2}	$\ldots$	s_1	
$z = 2$	d_1	d_0	s_{k-1}	$\ldots$	s_2	
$\vdots$						
$z = k$	d_{k-1}	d_{k-2}	$\ldots$	d_1	d_0	(destination)

Let $i_{k-1} i_{k-2} \ldots i_1 i_0$ be the address of the node under consideration. The following steps (executed by the node) describe the PATH.1 algorithm:

1. If the label of the node is the same as the destination address in the routing tag, then accept the message.

2. Otherwise, check the value of the routing tag counter z. The address of the next node in the path is $d_z i_{k-1} i_{k-2} \ldots i_1$ (route 1) or $i_{k-2} i_{k-3} \ldots i_1 i_0 d_{k-z-1}$ (route 2).

3. Increment the counter z and route the message to the next node.

Figure 7.10 shows a path from node 011 to node 101 in a DG(2, 3) network using route 2 of the PATH.1 algorithm.

Table 7.3 Route 2

$z = 0$	s_{k-1}	s_{k-2}	$\cdots$	s_1	s_0	(source)
$z = 1$	s_{k-2}	s_{k-3}	$\cdots$	s_0	d_{k-1}	
$z = 2$	s_{k-3}	$\cdots$		s_0	d_{k-1}	d_{k-2}
$\vdots$						
$z = k$	d_{k-1}	d_{k-2}	$\cdots$	d_1	d_0	(destination)

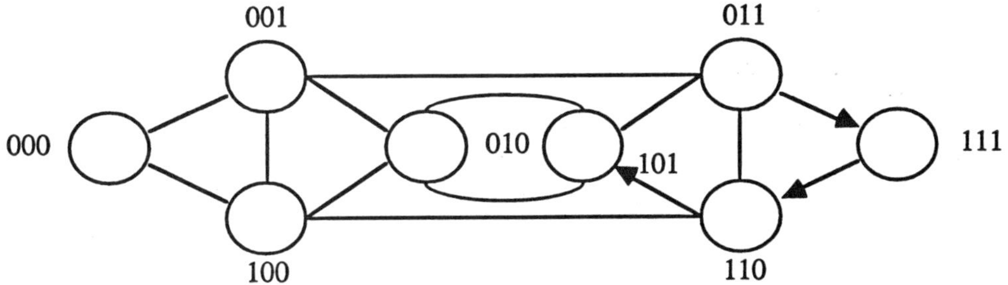

Figure 7.10 Routing using route 2 of PATH 1 algorithm. Source = 011, destination = 101, path is 011 to 111 to 110 to 101. Takes three steps.

7.7.5 PATH.2 Algorithm

The PATH.2 algorithm routes the message along the shortest path between the source and destination nodes. To find the shortest path, we treat the node addresses as binary strings and use a string matching algorithm described in [42]. Find the largest x such that

$$s_{x-1}s_{x-2}\ldots s_1 s_0 = d_{k-1}d_{k-2}\ldots d_{k-x}$$

and the largest y such that

$$s_{k-1}s_{k-2}\ldots s_{k-y} = d_{y-1}d_{y-2}\ldots d_1 d_0.$$

We can compute x and y in $O(k)$ [i.e., $O(\log N)$ time]. The following three cases arise depending on the relationship between x and y:

Case 1: $x > y$, the shortest path is given by the following sequence of nodes:

$$
\begin{array}{llllll}
z = 0 & s_{k-1} & s_{k-2} & \cdots & s_1 & s_0 & \text{(source)} \\
z = 1 & s_{k-2} & s_{k-3} & \cdots & s_0 & d_{k-x-1} & \\
z = 2 & s_{k-3} & \cdots & s_0 & d_{k-x-1} & d_{k-x-2} & \\
\vdots & & & & & & \\
z = k - x & s_{x-1} & s_{x-2} & \cdots & d_1 & d_0 & \text{(destination)}
\end{array}
$$

The destination $s_{x-1}s_{x-2}\ldots d_1 d_0 = d_{k-1}d_{k-2}\ldots d_1 d_0$ is reached after $k - x$ steps.

Case 2: $x < y$, the shortest path is given by the following sequence of nodes:

$$
\begin{array}{llllll}
z = 0 & s_{k-1} & s_{k-2} & \cdots & s_1 & s_0 & \text{(source)} \\
z = 1 & d_y & s_{k-1} & \cdots & s_2 & s_1 & \\
z = 2 & d_{y+1} & d_y & s_{k-1} & \cdots & s_2 & \\
\vdots & & & & & & \\
z = k - y & d_{k-1} & d_{k-2} & \cdots & s_{k-y+2} & s_{k-y+1} & \text{(destination)}
\end{array}
$$

The destination is reached after $k - y$ steps.

Case 3: $x = y$, choose either of the above routings to obtain the shortest path.

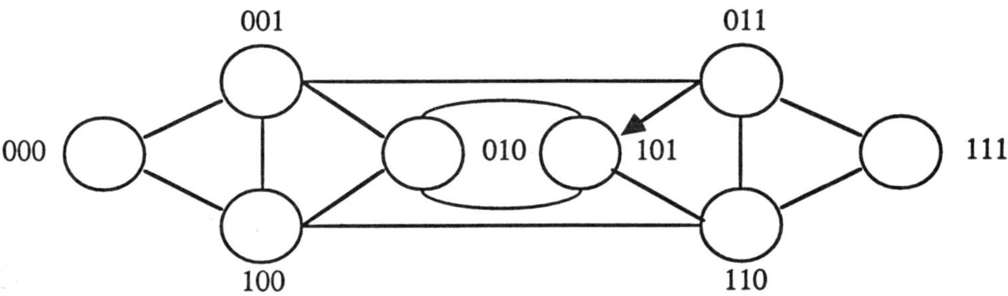

Figure 7.11 Example of shortest-path-routing. PATH 2 algorithm routes 011 to 101 using no hops.

Figure 7.11 shows the shortest path between nodes 011 and 101 in a DG(2, 3) network using the PATH.2 algorithm. In this case, $x = 1$ and $y = 2$; hence, the shortest path is of length 1. Using the PATH.1 algorithm yields a path length of 3.

The following steps describe the PATH.2 algorithm (as executed by node i):

1. If the label of the node $(i_{k-1}i_{k-2}\ldots i_1 i_0)$ is the same as the destination address in the routing tag, then accept the message.

2. Find the largest x such that $s_{x-1}s_{x-2}\ldots s_1 s_0 = d_{k-1}d_{k-2}\ldots d_{k-x}$ and the largest y such that $s_{k-1}s_{k-2}\ldots s_{k-y} = d_{y-1}d_{y-2}\ldots d_1 d_0$. If $x > y$ then the

address of the next node in the path is $i_{k-2}i_{k-3}\ldots i_1 i_0 d_{k-x-z}$, where z is the value of the counter. If $y > x$, then the address of the next node in the path is $i_{y+z-1}i_{k-1}i_{k-2}\ldots i_1$.

3. Increment the counter z and route the message to the next node whose address was generated in step 2.

Note that the value of x and y need not be computed by all nodes in the path. Instead, the value of x or y can be transmitted in the message header.

7.7.6   Routing between Layers

Let us now consider the problem of routing between layers. Let the source S and destination D nodes be at levels L and L - X, respectively. At the source, the interlevel routing (IRB) bit is set to 1 to indicate that the source and destination nodes are in different levels. Further more, when the IRB bit is set, the routing tag counter is not incremented in order to maintain a proper value of the counter for intralevel routing following the interlevel routing.

The source node S first routes the message to its parent str(S). This procedure is repeated recursively till the message is received by a node at the same level as the destination node D. The IRB bit is reset to 0 now, and the source address is replaced by the address of the node that received the message. The message can be then routed to the destination using the PATH.1 or PATH.2 algorithm.

When the destination is at a higher level than the source, routing can be similarly done by using app() to generate the address of the next node in the path till the message reaches the same level as that of the destination node. The message can then be routed using the PATH.1 or PATH.2 algorithm. Note that messages in the MBD are usually routed from higher to lower levels only.

7.7.7   Performance of MBD

Nearly all sensor networks used in *process control* industries are based on the *bus* or *ring* systems. The *bus* uses a common data path. The *ring* has *high* diameter and *low* connectivity. The *bus* and *ring* are not suitable to support the communication required in a large scale DSNs (especially using high sampling rates). Since *data integration* is by nature *hierarchical*, a hierarchical interconnection network would be most suitable for such a function.

Table 7.4 compares the topological, routing, and fault-tolerant properties of the MBD with the bus and ring networks. It can be seen that the MBD is a good alternative to the bus and ring networks.

7.8   Algorithmic Requirements

The abstract estimate is essentially a range of values within which we are sure the value of the physical variable resides. In the Flat Tree architecture, there are three

Table 7.4 Comparison of Networks

	Bus	Ring	MBD
Number of nodes	N	N	N
Diameter	$O(1)$	$O(N)$	$O(\log N)$
Degree	1	2	7
Routing	Simple	Simple	Simple
Fault-tolerance	1	2	$O(\log N)$
Cost (number of links)	1	$N - 1$	$< 3.5N$

types of nodes: leaf, intermediate, and root. With the exception of *root* nodes, each node functions as an *abstract sensor* to produce an *abstract estimate* to send to its parent node in the sensor cluster unit tree. This estimate may be either a one-interval abstract estimate (in the case of a *leaf* node) or a three-interval abstract estimate (in the case of an *internal* node).

7.8.1 Temporally Close Sensors

To determine if abstract sensor estimates are temporally close to each other, we need the following lemma and theorem.

Lemma 7 *Let a message be received by PE p at $C_p(t)$. Then this message was sent by PE p or its neighbors in the interval*

$$[C_p(t) - 2\epsilon - \delta_{max}, C_p(t) + 2\epsilon - \delta_{min}].$$

Proof: The proof is trivial if the message was sent by PE p's sensor since $C_p(t)$ lies in the interval.

Whenever a node sends a message, it time stamps the message with the current time. Because of the delay characteristics of the channel connecting PE p to its neighbor, the message was sent in the interval

$$[C_p(t) - \delta_{max}, C_p(t) - \delta_{min}]$$

according to p's clock. Let TS be the time stamp on the message. Then from equation(7.5), it follows that

$$TS \in [C_p(t) - 2\epsilon - \delta_{max}, C_p(t) + 2\epsilon - \delta_{min}].$$

The time stamp may not belong to the interval, say, if the channel becomes faulty. *Definition:* Let an abstract estimate time stamped at TS be received by PE p at time $C_p(t)$. The estimate is said to be *proper* if

$$TS \in [C_p(t) - 2\epsilon - \delta_{max}, C_p(t) + 2\epsilon - \delta_{min}]$$

Theorem 4 *Let the three proper abstract sensor estimates I_1, I_2 and I_3 be received by PE p at times $C_p(t_1) < C_p(t_2) < C_p(t_3)$ respectively Then I_i (i = 2, 3) can be integrated iff*

$$(C_p(t_i) - C_p(t_1) + 4\epsilon + \delta_{\max} - \delta_{\min}) \leq \gamma. \tag{7.11}$$

Proof. Since the estimates are proper estimates, we can deduce, using Lemma 7, that these estimates originated in the interval

$$[C_p(t_j) - 2\epsilon - \delta_{max}, C_p(t_j) + 2\epsilon - \delta_{min}]\,(1 \leq j \leq 3). \tag{7.12}$$

If I_2 is to be integrated with I_1, the spanning time interval of estimates I_1 and I_2 should at most equal γ, the maximum tolerance in time that a sensor can tolerate. The width of the interval that spans I_1 and I_2 is (obtained from 7.12 with $j = 1, 2$)

$$C_p(t_2) - C_p(t_1) + 4\epsilon + \delta_{\max} - \delta_{\min}.$$

Similar argument can be made for fusing I_3 with I_1. Inequality 7.11 then follows.

7.8.2 Sensor Integration in a Flat Tree

There are three distinct types of nodes to be considered in a sensor cluster unit tree: *leaf* nodes, *internal* nodes, and *root* nodes. The processing steps for each of these cases are different.

First consider a *leaf* node. A sensor/processor has to

- Compute an abstract estimate based on the value it obtains from its physical sensor

- Communicate this information directly to the sensor/processor above it in the tree

Next consider an *internal* node. A sensor/processor has to

- Compute an abstract estimate (as described for the leaf nodes)

- Combine this value with the abstract estimates received from its children before sending the information up the tree

Each sensor/processor does this in such a manner that there is no exponential growth of information. Such a growth would mean higher communication costs and higher computational cost at the root node.

Finally, consider a *root* node. A root node

- Performs the same functions as an internal node

- Acts as a control node, combining information from all the other sensors in the sensor cluster unit to produce a single output to be communicated to other commanders.

7.8.3 Sensor Integration in a MBD

The idea behind the process of sensor integration in the MBD is to

- keep the communication requirements small. This is done by communicating the abstract estimate as a single interval.

- Maintain accuracy by ensuring that the physical values of interest are always contained in the abstract estimate.

Since the binary de Bruijn network has a connectivity of 2, the MBD can tolerate at most one node fault or link fault per level (except at the topmost level, which is fully connected). We [27] give the results related to fault tolerance when abstract estimates (or intervals) are to be integrated in the presence of faults in the network.

Lemma 8 *Consider $N(N \geq 3)$ intervals of which at most one can be faulty. Then there can be at most two $(N - 1)$ distinct interval intersections among these N intervals.*

Proof: Without loss of generality, assume that the N intervals $(i_l, i_u)(1 \leq i \leq N)$ are sorted in *increasing* order by their *lower* bounds.

The *proof* is by *induction*. The *base* case for $N = 3$ is straightforward to prove by enumeration (see Fig. 7.12).

Consider p intervals and assume that the lemma holds for less than p intervals. Consider the first $(p-1)$ of the p sorted intervals. Since we know that there can be at most one fault, either all the $(p-1)$ intervals intersect or exactly $(p-2)$ intervals intersect.

Case 1: Exactly $(p - 2)$ intervals intersect (there is one fault among the $(p - 1)$ intervals): By the induction hypothesis, there are at most two $(p - 2)$ interval intersections, A (A_l, A_u) and B (B_l, B_u); let $A_l < B_l$. Further more, A and B are nonintersecting (i.e., $A_u < B_l$), otherwise, they would have formed a $(p - 1)$ intersecting interval. Since there can be one fault at most, the pth interval has to be correct and has to intersect with B, giving rise to one $(p - 1)$ intersecting interval.

Case 2: All $(p - 1)$ intervals intersect: Let the intersecting interval be C (C_l, C_u). By the induction hypothesis, there are at most two $(p - 2)$ interval intersections D (D_l, D_u) and E (E_l, E_u) and these overlap; that is, $D_l \leq E_l, E_l < D_u$. Further more, since C is the intersection of D and E, $C_u = \min(D_u, E_u)$ and $C_l = E_l = (p - 1)_l$, the lower bound of the $(p - 1)$st interval. Now the pth interval can intersect with some or all the intervals C, D, and E. Several cases arise:

Case 2(i): The pth interval (p_l, p_u) intersects C; the p intersecting interval is then $(p_l, \min(p_u, C_u))$, that is, $(p_l, \min(D_u, E_u, p_u))$. The pth interval could intersect either D (if $D_u > E_u$) or E (if $E_u > D_u$) but not both to form another $(p - 1)$ intersection.

Case 2 (ii): The pth interval (p_l, p_u) does not intersect C. Hence, $p_l > \min(D_u, E_u)$. The pth interval intersects either D (if $D_u > E_u$) or E (if $E_u > D_u$) but not both to form a $(p - 1)$ intersection.

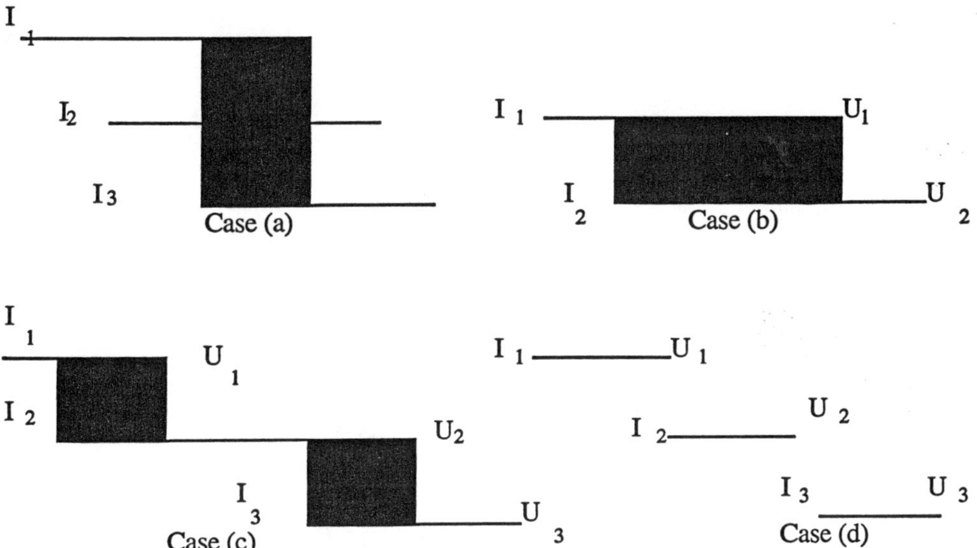

Figure 7.12 Ways in which three intervals can intersect.

The $(p-1)$ intersecting intervals arising from case 2 are the interval C and at most one more from cases 2(i) and 2(ii). The lemma then follows from the above cases. A direct consequence of Lemma 8 is the following theorem.

Theorem 5 *Given a set of N intervals containing at most one faulty interval,*

1. *There is no faulty interval if there is no $N-1$-interval intersection.*

2. *The interval not intersecting with an $N-1$-interval intersection is faulty if there is exactly one $N-1$-interval intersection.*

3. *There are two potentially faulty intervals if there are two $N-1$-interval intersections one of which is incorrect.*

In the third case, the two potentially faulty nodes can be traced by taking the *set difference* of the *interval names* that belong to each $(N-1)$ interval intersection.

Information integration process is done both within a level and between levels.[2] *Abstract estimates* move upward from the leaf nodes to the *commander* nodes. Every nonleaf node of the network combines the abstract estimates of its two children and the local sensor (sensor associated with this PE) to arrive at a new abstract estimate (AE^i). This step is called the *integration step*.

In the *integration step*, we assume that at most one of the three (local sensor and two children) received abstract sensor estimates is incorrect. The new abstract estimate is found from the three cases (refer Fig. 7.12) that could arise (Theorem 5).

If there are two interval intersections, then the smallest interval containing these intervals forms the new abstract estimate. It can also be shown [53] that this new estimate is at most as wide as one of the input abstract estimates.

Next, each node sends its AE^i to all its neighbors. When a node receives $AE^i s$ from its neighbors, it combines them to arrive at a new estimate AE^f. This step is called the *comparison* step, and the algorithm used to combine the estimates is similar to the one described for the *integration step*. In this step, however, a node combines three, four, or five estimates depending on the number of its neighbors (two, three, or four respectively).

Since the MBD can tolerate at most one fault (node or link) per level, one of the estimates received from a neighbor could be incorrect. Hence, when a node receives i ($i = 3$, 4 or 5) intervals in the comparison step, it chooses the smallest interval containing all (which is at most two as shown in Theorem 8) the $i-1$-interval intersections as the output. The width of this abstract estimate is again at most as wide as one of the input correct intervals.

Figure 7.13 shows the comparison process in a node of the network. If there are two $i-1$-interval intersections in the comparison step, we know that an incorrect interval exists. Identifying the faulty node that sent this incorrect interval requires the diagnostic testing of at most two nodes, as shown in Theorem 5. Once a node has been identified as faulty, appropriate action can be taken so as to either repair the faulty node or replace it and notify the parent and children of the faulty node. Here we shall not concern ourselves with the problems of identifying the cause of faulty behavior and attempting to rectify that cause.

The following steps summarize the process of information integration:

1. Receive abstract estimates from children and integrate them with the abstract estimate from a local sensor to get AE^i.

2. Send AE^i to neighbors.

3. Receive an abstract estimate from neighbors and compare with our abstract estimate to compute AE^f. Identify any faulty node in the process.

4. Send AE^f to the parent node.

---

[2] For convenience and brevity, the information integration of abstract estimates between distinct levels is referred to as integration. Likewise, the information integration within a level is referred to as comparison.

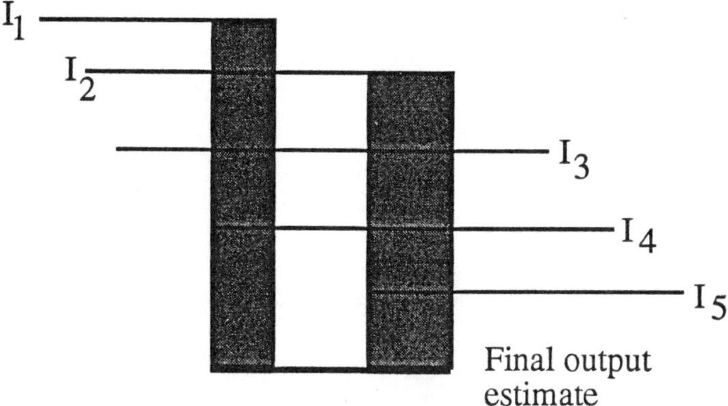

Figure 7.13 Comparison at a MBD node.

Note that this process of information integration ensures that only the correct estimates move up to the commander nodes in the network. Further more, the *width* of the estimates moving upward is *bounded* by the width of one of the correct estimates of that level of the MBD. An *incorrect* estimate would be received by a parent only when the child or the link connecting the two nodes is faulty.

Figure 7.14 shows the complete information integration process at a node in the network.

7.8.4 Fault Tolerance Issues in a MBD

In a large network it is unrealistic to expect all the nodes or links along a path to be fault free at all times. When some nodes or links fail, an alternative path that avoids the faulty node or link must be derived. One major advantage of a MBD network over the flat tree network is that abstract estimates can be routed around faults using the interconnections between nodes at the same level.

A node is faulty if it sends an incorrect abstract estimate to its parent or to any of its neighbors. Link faults can be detected if a node does not receive the abstract estimate of its neighbor during the comparison step. When a node (a node failure is assumed to be equivalent to the failure of all links associated with it) or link failure is detected, any of the following actions can be taken:

1. The fault can be ignored during the integration process. After integration is complete and abstract estimates have been sent to the upper level of the MBD, the node that detected this fault can run a diagnostic algorithm on the faulty node or link after isolating it.

2. If the faulty node or link is in the path of an abstract estimate transmitted

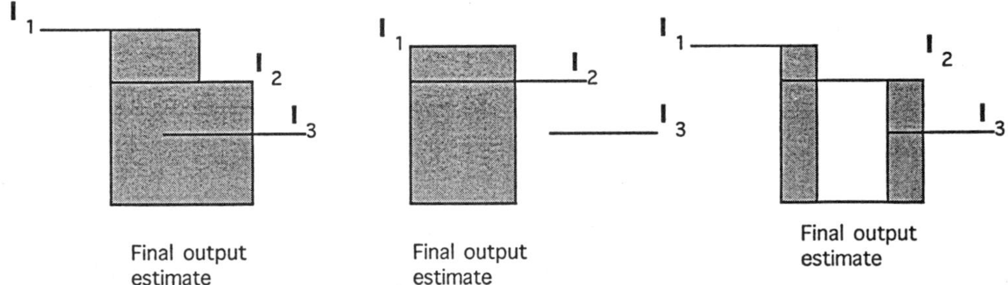

Figure 7.14 Information integration at a node.

toward its destination, this abstract estimate can be rerouted around the fault to the destination. After the integration process is complete, the node that detected the fault can run a diagnostic algorithm on the faulty component.

The MBD network provides fault tolerance by taking both of the remedial actions mentioned above. Suppose a node X, with children Y and Z, is faulty. In the *flat tree* network the subtree rooted at X is unusable. In the MBD network, however, the abstract estimates of Y and Z are also read by the neighbors of Y and Z. Thus, the abstract estimates of Y and Z get factored into the final abstract estimates produced by the neighbors of Y and Z. Hence, the subtree rooted at X does not become unusable; only the faulty node is unusable. Moreover, X is identified as a *faulty* node during the comparison step because its abstract estimate (which it sends to its neighbors) may not contain the physical value.

If action 2 is taken by the neighbor of the faulty node, then it must *reroute* the abstract estimate received around the faulty node to the destination. This means that the destination node must wait for more time to receive the abstract estimate, because additional hops may be required for rerouting the message. This requires that the value of γ (maximum difference in time that a node can tolerate between intervals that can be integrated) be increased to maintain the near synchronous behavior of the sensor network. Note that by increasing the value of γ the network would tolerate a single node or link fault, but the process of sensor integration would be slowed down.

Samantham and Pradhan [68] mention that four additional hops are enough to avoid a single node fault in a binary de Bruijn network. Since the nodes in every level (except the top level) in the MBD are arranged in a binary de Bruijn network, the value of γ will have to be increased by four time units.

We show one way of avoiding a single node fault using exactly four hops when routing in any level of the MBD except the topmost level. Let $s_{k-1}s_{k-2}\ldots s_1 s_0$ be the source node and $d_{k-1}d_{k-2}\ldots d_1 d_0$ be the destination node. Application of the PATH.1 algorithm yields the following path:

Table 7.5 Path given by PATH.1 Algorithm

$z=0$	s_{k-1}	s_{k-2}	$\ldots$	$\ldots$	$\ldots$	s_1	s_0	(source)
$z=1$	s_{k-2}	s_{k-3}	$\ldots$	$\ldots$	$\ldots$	s_0	d_{k-1}	
$z=2$	s_{k-3}	$\ldots$	$\ldots$	$\ldots$	s_0	d_{k-1}	d_{k-2}	
$z=m-1$	s_{k-m}	s_{k-m-1}	$\ldots$	s_0	d_{k-1}	$\ldots$	d_{k-m+1}	i_1
$z=m$	s_{k-m-1}	s_{k-m-2}	$\ldots$	d_{k-1}	$\ldots$	d_{k-m+1}	d_{k-m}	i_2
$z=m+1$	s_{k-m-2}	$\ldots$	d_{k-1}	$\ldots$	$\ldots$	d_{k-m}	d_{k-m-1}	i_3
$\vdots$								
$z=k$	d_{k-1}	d_{k-2}	$\ldots$	$\ldots$	$\ldots$	d_1	d_0	(destination)

Assume that either node i_2 or the link between i_1 and i_2 has failed. We now show an alternative route (reroute) between i_1 (rerouting source) and i_3 (rerouting destination) that takes only four additional hops. Note that this technique is independent of the number of nodes in the level.

When i_1 receives a message (consisting of the abstract estimate and the Type 1 tag), it appends four fields to the Type 1 tag that enable rerouting of the message:

1. Source address (i_1)

2. Destination address (i_2)

3. Reroute counter (RRC)

4. Rerouting bit (RRB)

We shall refer to the tag formed by appending reroute fields to the Type 1 tag, as the Type 2 tag. Figure 7.15 shows a Type 2 tag. To initiate rerouting, i_1 increments z and sets RRB = 1. When RRB = 1, a node does not increment z; instead it uses RRC to compute the address of the next node in the reroute. When the message reaches i_2, i_2 removes the reroute fields from the tag and increments z. Routing from i_2 then proceeds normally using the PATH.1 or PATH.2 algorithm.

The alternative route between i_1 and i_2 is shown in Table 7.6.

This route takes six steps only four more than the normal route between i_1 and i_2. Figure 7.16 shows fault-tolerant routing in a DG(2, 3) (level 3) between nodes

Type 1 tag	Source address	Destination address	Reroute counter RRC	Rerouting bit RRB

Figure 7.15 Type 2 Tag consists of type, IRB, source address, destination address, routing tag counter, new source address (i_1), new destination address (i_2), reroute counter, and rerouting bit.

Table 7.6 Alternative Route

$z = m - 1$	s_{k-m}	s_{k-m-1}	$\cdots$	s_0	d_{k-1}	$\cdots$	d_{k-m+1}		
$z = m$	s_{k-m-1}	s_{k-m-2}	$\cdots$	s_0	d_{k-1}	$\cdots$	d_{k-m+1}	$\bar{d}_{k-m}$	
$z = m$	s_{k-m-2}	s_{k-m-3}	$\cdots$	s_0	d_{k-1}	$\cdots$	d_{k-m+1}	$\bar{d}_{(k-m)}$	d_{k-m-1}
$z = m$	d_{k-m-1}	s_{k-m-2}	$\cdots$	s_0	d_{k-1}	$\cdots$	d_{k-m+1}	$\bar{d}_{k-m}$	
$z = m$	d_{k-m}	d_{k-m-1}	$\cdots$	s_0	d_{k-1}	$\cdots$	d_{k-m+1}		
$z = m$	d_{k-m-1}	s_{k-m-2}	$\cdots$	s_0	d_{k-1}	$\cdots$	d_{k-m+1}	d_{k-m}	
$z = m + 1$	s_{k-m-2}	s_{k-m-3}	$\cdots$	s_0	d_{k-1}	$\cdots$	d_{k-m+1}	d_{k-m}	d_{k-m-1}

001 and 110 when the node 011 is faulty. This alternative route can be chosen when a faulty node is encountered in the path to the destination node. Hence, the routing algorithms given earlier can be easily adapted to take the alternative path in case of faults. Our rerouting algorithm is also more adaptive to faults than the one presented in [59], since the former does not require that the presence of a fault be known to the source node as the latter does.

Finally, since the network can sustain one node or link fault at every level, the MBD network with L levels and $N = 2(2^L - 1)$ nodes can sustain L, that is, approximately $\log N$, node or link faults.

Some Important Issues in a DSN

Some of the important issues in a DSN that need to be examined are the following.

1. How should the observations available at a sensor or a sensor cluster be processed so as to recover the relevant signal parameters?

2. What are the message routing strategies for the network?

3. How does internodal coupling affect obtaining unbiased estimates of a param-

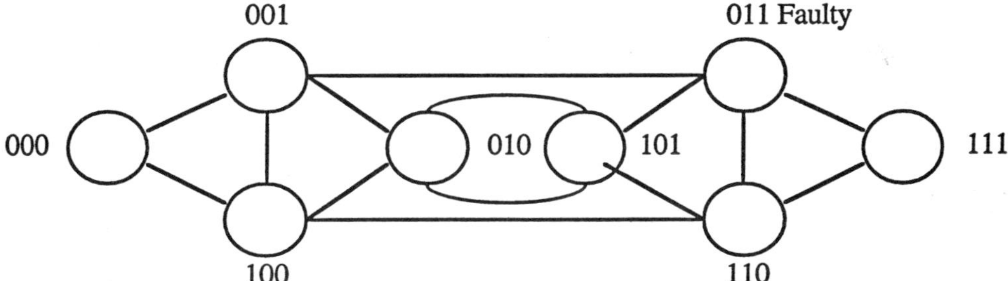

Figure 7.16 Rerouting after a fault. Fault-tolerant routing, between 001 and 110 when 011 is faulty.

eter?

4. How do we structure the signal processing cluster units?

The other issues of importance related to message passing, which are not examined in detail here, are 1) fail-safe message passing strategies and 2) message routing strategies when multiple paths exist between two nodes in the DSN.

To illustrate a problem relating to internodal coupling, consider Fig. 7.17. The estimate that P receives from L could itself be a function of P's estimate since, in addition to the path from L to P, there exists a path from P to L. As a result, the combined estimate calculated at p could be a biased estimate favoring P's estimate since the latter has been included twice in the calculation for the combined estimate. One reason we choose the binary tree architecture is that such a problem does not arise in these networks.

7.8.5 Information Integration Algorithm

The integrated information is a k-interval abstract sensor I_p computed from I_1, I_2, I_3. Thus, $I_p = D(I_1, I_2, I_3)$. The computation of the function D is the information integration algorithm. As we have mentioned earlier, we consider only a one-interval information integration algorithm. Furthermore, we assume that the state of the system at any instant can be represented by a single value. (The algorithm can be extended in a straightforward manner for the case of a vector of values with each value being represented by a k-interval estimate.)

The algorithm, One_Interval_Integrate, shown in pseudo-Pascal, computes the one-interval abstract estimate at a node p and sends it to its parent time stamping it

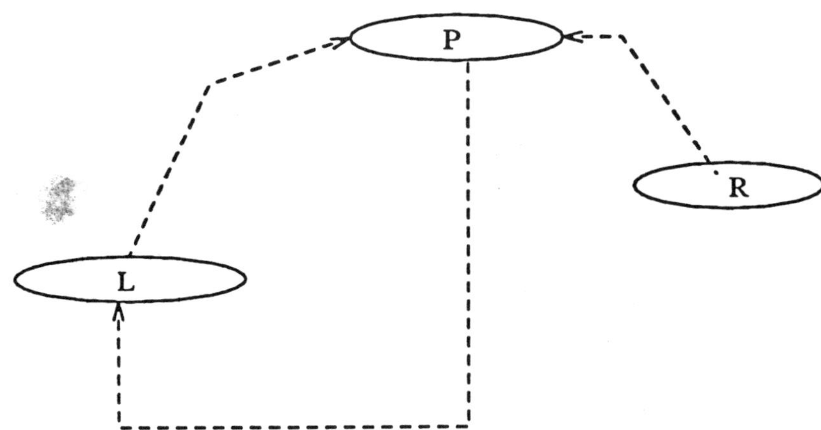

L's integrated sensor value that P obtains could be a function

of P's value since there exists a path from P to L

Figure 7.17 Internodal coupling. $L's$ integrted sensor value that S obtains could itself be a function of $P's$ value since there exist's a path from P to L.

with the current time CT. Procedure FORM_I1_I2_I3, which is repeatedly called as a procedure by One-Interval-Integrate, integrates the one-interval interval estimates that are temporally close to each other.

Algorithm

One_Interval_integrate;
    /* Δ = minimum width of overlap for two intervals considered intersecting. */
    /* CT = current time as seen by a PE clock. */
    /* I_p = integrated abstract sensor estimate. */

begin FORM_I1_I2_I3
        if width$(I_1 \cap I_2 \cap I_3) \geq \Delta$
        then /*
            l_1 := lowest value in the intersecting interval;
            h_1 := highest value in the intersecting interval;
        else if $(\text{width}(I_i \cap I_j) \geq \Delta)$ **and**
           $(\text{width}(I_j \cap I_k) \geq \Delta)$
               then $(i \neq j, j \neq k, i \neq k, 1 \leq i, j, k \leq 3)$
               l_1 := lowest value in the left intersecting interval;
               h_1 := highest value in the right intersecting interval;
               else if $(\text{width}(I_i \cap I_j) \geq \Delta)$
                    then $(i \neq j, 1 \leq i, j \leq 3)$

$l_1 :=$ lowest value in the intersecting interval;
$h_1 :=$ highest value in the intersecting interval;
else /* no intersecting intervals */
$l_1 :=$ lowest value in the intersecting interval;
$h_1 :=$ highest value in the spanning interval;
endif
endif
endif
$I_p := [l_1, h_1];$
if p is not the root
then
send(parent(p), I_p, CT);
endif
end.

Procedure FORM_I1_I2_I3
/* Computes K-interval abstract sensors */
/* Buf = FIFO mesage buffer */
/* message = triple $\langle abstractsensor, timestamp \rangle$ */
/* messages are sent by the abstract sensors and received by the PE p */
/* empty(buf) is true when there is no message in the buffer */
/* $CT =$ current time (for each PE) */
begin
done := false;
while not done **do**
receive(I_1, t);
if $t \in [CT - 2\epsilon - \delta_{max}, CT + 2\epsilon - \delta_{min}]$
then
done := true;
/* I_1 is a proper K-interval abstract sensor */
first_sensor_time := CT;
/* store the time at which first proper K-interval abstract sensor read */
endif
enddo
$I_2, I_3 := \emptyset;$
if p is the leaf PE
then
goto L3;
/* exit a leaf PE that does not have any children */
endif
L1:   empty_2_3 := false;
while empty(buf) **do**
/* if current time exceeds the max transmission delay d then set $I_2, I_3 \leftarrow \emptyset$ */
if $((CT\_first\_sensor) + 4\epsilon + \delta_{max} - \delta_{min}) \geq \gamma$

```
        then
           I_2 := ∅;
           I_3 := ∅;
           empty_2_3 := true;
        endif
        if not(empty_2_3 )
        then /* a K-interval abstract sensor was read */
           receive(I_2, t);
           if not (t ∈ [CT − 2ε − δ_max, CT + 2ε − δ_min)]
           then /* i.e., this abstract sensor is not a proper abstract sensor */
           go to L1;
        endif
L2:     empty_3 := false;
        /* the second proper abstract sensor has been read. Wait for the third one */
        while empty(buf ) do
           if ((CT_first_sensor) + 4ε + δ_max − δ_min) > γ
           then
              I_3 := ∅;
              empty_3 := true;
           endif
        endwhile
        if not(empty_3)
        then /* third K-interval abstract sensor was read */
           receive(I_3, t);
           if not (t ∈ [CT − 2ε − δ_max, CT + 2ε − δ_min])
           then /* i.e., this abstract sensor is not a proper abstract sensor */
              go to L2;
           endif
           go to L3;
           endif
        endif
        endwhile
endprocedure.
```

Observe that in our integration algorithm the abstract sensor estimate associated with a PE at level i could affect the information integration in the same manner as the aggregate information produced by the children of the node rooted at level i. This method of integration could be altered by assigning weights to the intervals such that the individual sensor's abstract estimate is not as important as the abstract estimate produced by the aggregate. A sensor's abstract estimate could be weighted, for example, in inverse proportion to its level in the hierarchy.

7.8.6   Communication Constraints

The performance of any coordinated computing system depends on the communication and computational speeds. Communication poses a major problem since the time needed for communication is large in large networks. In a DSN, the sensors generate data repetitively at regular intervals. It is therefore essential to ensure that each datum is delivered to the destination node in finite time, before the data of the next cycle arrive at the node. Hence there is an upper bound on the maximum allowable transmission time in the network. Also, considering the environment that a DSN operates in, it is necessary to examine fail-safety requirements.

7.8.7   Delay Constraints

When a DSN node needs to transmit information to other nodes in the network, the communication should be completed before another set of data of the next cycle arrives at the node. The problem could be stated slightly differently. A node n_i needs to transmit a unit of information, to other nodes within time t_d, where t_d, is the maximum permissible transmission delay. The question is whether this is possible in the present configuration of the network. From time to time this check needs to be done in the network for proper functioning of the DSN.

The factors that affect transmission delay in the DSN include the capacity of the channel, number of lines on the route, propagation delay, and packet size. From queuing theory results [7], we have an approximation for the mean total delay T:

$$T = \frac{L}{C(1-\rho)} + t_p$$

where L = mean length of frames, C = channel capacity, ρ = load on the line, and t_p = propagation delay.

For a network, the parameters L, C and ρ are generally known to an acceptable degree of approximation. For a dynamic network, it is necessary to determine t_p. The propagation delay depends mainly on the length of the route, and hence the maximum end-to-end distance is the worst case measure of the propagation delay.

A network can be represented as a graph $G(V, E)$ where V is the set of vertices in the graph denoting the nodes in the network, and E is the set of edges in the graph denoting the communication links of the network.

The following are generally assumed:

- The links are bidirectional.

- The network does not experience any failure during the execution of the algorithm.

- The messages are delivered to the other nodes in finite time.

- There is no loss of messages.

Let $d(u, v) =$ the length of the shortest path between u and v; $u, v \in V(G)$. The *diameter* $D(G)$ of a connected graph G is the *length* of the longest $d(u, v)$, for all $u, v \in V(G)$.

$$D(G) = \max_{u, v \in V(G)} d(u, v)$$

Thus, the diameter of the graph underlying the network gives a measure of the maximum end-to-end transmission delay.

We will develop a family of distributed algorithms to determine the diameter of the graph underlying a network.

7.8.8   Reliability Constraints

A vital consideration for automatic routing and proper functioning of DSN is the survivability of the network. Nodes and links in DSN may fail due to several reasons, considering the fact that a DSN works in hostile environments. It is therefore necessary to maintain the message flow in the network.

When a node or link fails, it is necessary to find alternative routes to route messages. Distributed protocols for such dynamic routing are discussed in the next section. However, it is necessary that the topological configuration of the network have sufficient connectivity such that when some links and/or nodes fail, the network remains connected. This requirement should be taken into consideration while designing network topologies. Suppose at most L links fail in the network. Then the graph underlying the network must be $(L + 1)$ edge connected for the network to remain connected despite the failure of L links. Similarly, for the network to remain connected if k nodes fail, the graph must be $(k + 1)$ connected. Thus, it is possible to keep the network connected despite link or node failures by increasing the connectivity of the graph. However, a node or link failure can increase the diameter of the graph, thus increasing the end-to-end delay in the network.

Suppose $D(G)$ is the diameter of the graph G underlying a network. We assume that we are given a k-connected graph of order n, and that there will be at most m node failures at any instant of time in the network. Let us now derive a condition for k such that under m node failures the diameter of the network does not exceed the maximum permissible diameter D', for which the transmission delay does not exceed t_p. In [7], it is shown that after deletion of m nodes in a k-connected graph of order n $(m < k)$, the diameter of the resulting graph D' is bounded by

$$D' \leq \frac{n - m - 2}{k - m} + 1 \ .$$

It is clear that for D' to be within the delay constraint, under a fixed m, the connectivity k of the graph could now be computed using the above inequality, such that not only network connectedness is maintained, but also the node failures do not result in unacceptable delays. Also, for an l edge-connected graph of order n, deletion of $l - 1$ results in a graph of diameter D' such that

$$D' \leq \Gamma D + \Gamma - 1$$

where D is the diameter of the original graph [7].

We have shown the effect of link or node failures on the connectedness of the graph and the delay in the network. These results provide means to retain the network connection, and also to ensure that the delay constraints for the DSN are met.

7.9 Diameter of a Graph

A *graph* $G(V(G), E(G))$ consists of $V(G)$, a set of *vertices* and E, a set of *edges*. We may also write $G(V, E)$, V and E. A *network* can be represented as a graph where V, the set of vertices, denotes the *nodes* in the network, and E, the set of edges, denotes the *communication links* of the network. The *length* of a path between two nodes is the number of edges encountered while going from one node to the other. The *distance* between two nodes is the shortest length between the nodes.

The *diameter* of the network is the largest distance between any two nodes in the network.

Let $d(u, v) =$ the length of the shortest path between u and v; $u, v \in V(G)$.

The *diameter* $D(G)$ of a connected graph G is the *length* of the longest $d(u, v)$, for all $u, v \in V(G)$.

$$D(G) = \max_{u,v \in V(G)} d(u, v) \, .$$

The *degree of a node* is the number of edges associated with that node. The *degree of the network* is the largest degree of any node in the network. For more details on graph theory, see, for instance, [23].

We will develop two distributed algorithms for finding the diameter of an asynchronous tree network:

- SDIA: single node starts the algorithm

- MDIA: multiple nodes start the algorithm

The algorithms can be easily extended to general networks. Note that the diameter can also be determined using the center-finding algorithm in [43], but such algorithms do not support multiple start nodes and are also not optimal for finding the diameter.

7.9.1 Distributed Diameter-finding Algorithms

The idea behind the SDIA and MDIA algorithms is as follows:

- Determine two largest heights of each node, say h_1 and h_2

- Determine the node with the largest $h_1 + h_2$

- Determine the diameter of the tree.

If the start node lies on the diameter path, then it computes the diameter; otherwise, it receives the diameter information from a node that computed it.

Two messages are used in the algorithms.

find This message has only the message header.

return This message has two parameters.

> 1. d is the the maximum height of the subtree rooted at the sender.

> 2. p_d is the maximum end-to-end distance at the sender.

7.9.2 Algorithm SDIA

SDIA assumes that a single node, say s, starts the algorithm. Let $T(V(T), E(T))$ be the tree of order n (i.e., $|V(T)| = n$). We can view the tree T with s as its root and having subtrees $T_1, T_2, \ldots, T_l (1 \le l \le n - 1)$. that is, $T_s = \{T_1, T_2, \ldots, T_{n-1}\}$. It is clear that $V(T_i) \cap V(T_j) = \{s\}, (i \ne j)$.

The algorithm is as follows. All nodes are initially in *idle* state. The start node s broadcasts the *find* message to its neighbors and changes to the *wait* state. When an *idle* node receives the *find* message,

1. If it is a leaf node, then send $return(0, 0)$; quit.

2. An internal node broadcasts the message to all its neighbors except the one from which it received the message and transits to the *wait* state.

When a node in *wait* state receives the $return(d, p_d)$ message from a neighbor,

1. increment the count d of the message and compute the two largest d's received so far. Compute the maximum of the p_d's received.
and

2. *If* it has received the *return* message from all its neighbors in response to the *find* message, then send $return$ $(\max(h_1, h_2), \max(p_d, h_1 + h_2))$ to the neighbor from which it received the *find* message exit; else, remain in the same state.

The algorithm terminates when the start node s has received the *return* message from all its neighbors.

Properties

Suppose a node s starts the algorithm. Then

1. A node changes its state exactly once. Once from *idle* state to *wait* state and from *wait* state, it exits the algorithm.

2. A node $(\in s)$ in *wait* state does not receive a *find* message from any of its neighbors.

Proof: Suppose a node n_i in *wait* state receives a *find* message from its neighbor n_j; $n_i, n_j \in V(T), i \ne j$. Let n_k be the node that sent *find* message to an *idle* n_i.

Since s is the only initiator of the algorithm, there exist two paths from s to n_i: one path via n_j and the other via n_k. This is contradiction to the assumption that T is a tree.

Lemma 9 *The algorithm SDIA correctly determines the diameter of the tree and terminates in finite time.*

Proof: The *find* message, initiated at some start node s traverses down all the subtrees rooted at s to each leaf. Each internal node n_i computes the maximum end-to-end distance of the subtree rooted at n_i in T_s (p_d of the *return* message). Since the diameter of the graph is the maximum end-to-end distance, some internal node computes it. If s is on the diameter path, then s computes it, otherwise, s receives the computed diameter information from a node on the diameter path through its neighbor.

Since the *return* message moves up the tree starting from the leaf, the start node receives the *return* message from all its neighbors, thus terminating the algorithm in finite time.

Let $h(T_i)$ denote the height of the tree T_i, and

$$h_{\max} = \max\{h(T_j) | \forall T_j \in T_s\} \ .$$

Theorem 6 *Suppose a node s starts the algorithm SDIA. Then s determines the diameter of the tree T in at most $2h_{max}$ units of time, if each message is delivered over a link in at most one unit of time.*

Proof: The message *find* moves down the tree rooted at s and the *return* message retraces the path of the *find* message. Since each internal node broadcasts the *find* message, the maximum time it takes for the *find* message to reach a leaf is the timeto the one at the farthest distance from s. Hence, the total time taken for finding the diameter is $2h(T_i)$, where T_i is the subtree at s, which has the maximum height.

Theorem 7 *The algorithm SDIA finds the diameter of the tree T using exactly $2(n-1)$ messages and is optimal in message complexity.*

Proof: It is clear from the algorithm SDIA that the *find* message travels on all the edges of the tree, and the *return* message retraces the *find* message. Thus, there are two messages on each edge, and hence the algorithm uses exactly $2(n-1)$ messages.

For any *diameter-finding* algorithm, the message has to be sent over each link of the tree and hence the problem of finding the diameter has an obvious message complexity lower bound of $O(n)$. The algorithm is hence message optimal.

7.9.3 Algorithm MDIA

This is a generalization of the SDIA. The algorithm allows multiple nodes to starts the process of finding the diameter.

Suppose a set of nodes $S = \{s_1, s_2, \ldots, s_k\}, S \subseteq V(T)$, start the process of finding the diameter. The algorithm proceeds as follows.

Each start node s_i sends the *find* message to all its neighbors. If s_i is a leaf, it behaves as if it has received a *find* message and sends the *return* message to its neighbor.

When a node n_i in *idle* state receives a *find* message,

if n_i is a leaf, then change state to *active* send $return(0, 0)$; otherwise, change state to *active* broadcast *find* message to all its other neighbors.

When a node n_i in *idle* state receives a $return(d, p_d)$ message;

Can happen when some leaf starts the algorithm. *

if it has received *return* from all its neighbors except one n_j, say, then compute (d, p_d) from the received messages and send the $return(d, p_d)$ to n_j and set state to *active.*

if it has received *return* from all its neighbors, then transit to the *terminal* state and compute p_d from all the received messages and broadcast the SETDIA(p_d) message to all its neighbors; exit. Otherwise, compute d_1, d_2 as the two largest d's received and the maximum of the received p_d's; change to *active* state and broadcast *find* message to all other neighbors.

When a node in *active* state receives a *find* message, the message is ignored. It implies that some other node/s started the algorithm in the subtree. *

When a node n_i in *active* state receives a $return(d, p_d)$ from a neighbor,

If the *return* message is received from all its neighbors, then change state to *terminal;* broadcast SETDIA$(\max(d_1 + d_2), \max(p_d))$ message to all neighbors and exit.

If the *return* message is received from all neighbors except one, n_j, then $return(d, p_d)$ id sent to n_j. Otherwise, computes d_1, d_2 and p_d.

When a start node s_i receives a SETDIA(dia) message, it records the diameter and broadcasts the received message to other neighbors, and exits the algorithm.

The algorithm terminates after each node receives the SETDIA message.

Lemma 10 *Suppose a set of nodes* $S = \{s_1, s_2, \ldots, s_l\}$, $S \subseteq V(T)$, *starts the algorithm. There will be exactly one node or two adjacent nodes that reach the terminal state in finite time.*

Proof: A node reaches the *terminal* state whenever it has received the *return* message from all its neighbors. Also, a node sends a *return* message to its neighbor if and only if it has received a *return* from all its neighbors except one or if it is a leaf. So, if a node n_i, receives *return* from a neighbor n_j, it is implied that the necessary computations are completed with the subtree rooted at n_j.

Suppose there are three nodes n_i, n_j, and n_k $i \leq j \leq k$ (not necessarily adjacent) that a reach the *terminal* state. This implies that a *return* message was sent by n_i to the subtrees that include n_j and n_k; and by n_j to n_i and n_k, and by n_k to n_i and n_j, a contradiction to the facts stated above.

It is clear that there could be one node that reaches the *terminal* state. If two nodes n_i and n_j both reach the *terminal* state, both must have sent the *return* message to each other and hence must be adjacent.

Theorem 8 *The algorithm MDIA correctly finds the diameter of the tree in finite time and terminates in finite time.*

Proof: The *find* message is initiated by all the start nodes $S = \{s_1, s_2, \ldots, s_l\}$, $S \subseteq V(T)$. Each node $s_i \in S$ functions as if it is the only start node. A node sends the *return* message only when it has received *return* from all its neighbors except one. Hence, when a node receives the *return* message from its neighbor, it is implied that all the necessary computations are completed in the subtree rooted at the neighbor. Each internal node determines the maximum end-to-end distance in the subtree. Some node n_j will eventually receive the *return* message from all its neighbors and hence compute the diameter of the tree. All the nodes terminate the algorithm after receiving the SETDIA message, and hence the algorithm terminates in finite time.

Let $h(T)$ denote the height of the tree T and $D(T)$ the diameter of the tree. Let

$$h_{\max} = \max\{h(T_i)|\forall i \in S\}.$$

Theorem 9 *Suppose nodes $S = \{s_1, s_2, \ldots, s_l\}$, $1 \leq l \leq n$ and $S \subseteq V(T)$ start the algorithm MDIA. Then, each will determine the diameter of the tree in less than $2h_{\max} + D(T)$ units of time if each message is delivered in one unit of time.*

Proof: The maximum distance traversed by the *find* message originating at a start node s_i is equal to the maximum height of the subtree rooted at s_i. The *return* message retraces this path, and these two messages together account for the first term in the time complexity. Once a node reaches the *terminal* state, it broadcasts the SETDIA message to all other nodes in the network, which traverses less than the diameter path length that accounts for the second term in the time complexity relation.

Theorem 10 *The algorithm MDIA finds the diameter of the tree using at most $4(n-1)$ messages.*

Proof: In the worst case, all nodes in the tree might start the algorithm simultaneously and each would send the *find* message to their neighbors, except the leaf nodes which send the *return* message. This amounts to $2(n-1)$ messages. There will be $(n-1)$ *return* messages before a node can get to the *terminal* state. A node in *terminal* state broadcasts the SETDIA message which amounts to $(n-1)$ messages, and hence the message complexity of the algorithm MDIA is at most $4(n-1)$.

Finding the diameter distributively of an arbitrary graph is expensive in terms of communication complexity. The algorithm to determine the center of a graph in [43] could be employed to determine the diameter of an arbitrary graph. The algorithm is outlined as follows.

- Construct a spanning tree of the graph rooted at a node s using any of the existing algorithms [71].

- Each node in the tree then initiates a shortest-path-finding algorithm [44] and determines its eccentricity.

- Each node then computes its maximum end-to-end distance and sends this information to the root.

- The root node upon receiving the end-to-end distance information from each nodes in the network, computes the largest of these which is the diameter of the graph. Then the diameter information is broadcast to all other nodes in the network.

The communication complexity of the algorithm is computed as follows. The first step requires $O(|E|)$ messages, and the shortest-path-finding algorithm initiated at each node requires $O(|E|)$ messages. Thus, the total number of messages required to compute the shortest paths by all nodes is $O(n|E|)$. $O(n^2)$ messages are needed to perform the third step. The last step of the algorithm requires $O(n)$ messages. Hence, the message complexity is $O(n|E|)$, which in the worst case is $O(n^3)$.

7.10 Chapter Summary

In this chapter, we have considered the architectural issues involved in distributed sensor integration. The clock synchronization issue is dealt with within the DSN framework. Two architectures that are amenable for distributed sensor integration are proposed. They are the flat tree architecture and the multilevel binary de Bruijn network.

7.11 References and Further Reading

The material is adapted from [27], [28], [31], [35], [71].

Wesson et al. [80] were the first to discuss hierarchical and committee organizations in the context of distributed situation assessment.

Pradan [56, 57, 58] discusses several aspects of de Bruijn networks: (1) their use as interconnection topologies for fault-tolerant parallel and distributed architectures, (2) their use for VLSI architectures, and (3) fault-tolerant VLSI architectures.

Problem Set 7

7-1. What is a completely connected graph?

7-2. Give a formal description of a flat tree. Assume that each sensor cluster unit is a tree.

7-3. Define the following: abstract sensor, interval, width of interval, intersection of intervals, interval inclusion, and distinct intervals.

7-4. What are the properties of a de Bruijn network?

Chapter 8

Characterization of Sensor Integration

We have discussed several sensor integration techniques in the previous chapters. In this chapter, we will try to give a mathematical framework for sensor integration. We will do it in two steps.

First, we will describe a formal way to view one-dimensional sensor integation algorithms. As examples, we will cover algorithm M1 (proposed by Marzullo) and algorithm PIKM (proposed by Prasad, Iyengar, Kashyap, and Madan). Next, we will try to find a meaningful abstraction of a distributed sensor networks.

8.1 Functional Characterization of Sensor Integration

A concrete sensor is a device for sensing a physical variable in an environment. Examples of concrete sensors are thermometers, barometers, and radars. A concrete sensor returns a result, which can be a single number. However, since a sensor is prone to inaccuracies and there may be some uncertainty in the value of its output, Marzullo [53] proposed to represent sensor outputs as connected intervals on the real line R. No assumptions are made about the width of these intervals or their position on the real line.

An abstract sensor is a mathematical abstraction of the output given by a concrete sensor. For our discussion, it is an interval, say $I = (l, u)$ on the real line. l is the lower bound and u is the upper bound of the interval I associated with the sensor s.

Let v be the actual value. We say that the sensor s is correct if v is contained in I. In set theoeretic terms, s is correct if $v \in I$ (i.e., v is a member of I).

If there are several sensors, say N, then we will denote the sensors by $s_1, s_2, \ldots, s_N$. The corresponding intervals are $I_1, I_2, \ldots, I_N$. The jth interval is given by $I_j = (l_j, u_j)$ where $1 \le j \le N$.

A function that defines a membership of a set is called a characteristic function. The characteristic function χ_j of the jth sensor s_j $(1 \le j \le N)$ is given by

$$\chi_j : R \longmapsto \{0, 1\}$$

$$\chi_j(x) = \begin{cases} 1 & \text{if } x \in I_i \\ 0 & \text{otherwise} \end{cases}$$

$(1 \leq j \leq N)$.

A point x belongs to the interval I_j if and only if $\chi_j(x) = 1$. Likewise, x belongs to the interval I_k if and only if $\chi_k(x) = 1$.

From this, we can deduce that x belongs to both intervals I_j and I_k if and only if $\chi_j(x) = 1$ and $\chi_k(x) = 1$. This is the same as saying $\chi_j(x) \cdot \chi_k(x) = 1$, which means that x belongs to $I_j \cap I_k$ if and only if $\chi_j(x) \cdot \chi_k(x) = 1$. This can be generalized as "x belongs to the set intersection of the intervals if and only if the product of the corresponding characteristic functions is 1."

We can also represent other operations (e.g., union of two intervals) in terms of the characteristic functions.

8.1.1   Characterizing Marzullo's Algorithm

Sensor integration is the process of combining sensor readings in a meaningful manner. For instance, Marzullo's algorithm M1 shows a way to perform fault-tolerant integration of n redundant abstract sensors, when at most F of them can be faulty. M1 combines the I_j's $(1 \leq j \leq N)$ to obtain an abstract interval estimate $I_p = [a_p, b_p]$.

I_p should satisfy two properties. First, it should be a reliable and fairly accurate estimate of the region in which the physical sensor value lies. Second, the reliability of the integration process should not be severely affected by the bounded number of faulty sensors.

We can characterize algorithm M1 as follows: Obtain a functional relationship f between the characteristic function χ_p of $I_p = [a_p, b_p]$ and the $\chi_j (1 \leq j \leq N)$ such that $\chi_p(x) = f(\chi_1(x), \ldots, \chi_n(x))$ and $\chi_{p-1}(1)$ is a fault-tolerant interval estimate of the physical value being measured.

To express the functional f, we introduce several relevant operations and functions. For a real-valued function f, the *norm* of f is given by

$$\|f\| = \sup\{|f(x)| \mid x \in R\} \ .$$

That is, $\|f\|$ is the smallest real-number α such that $f(x) \leq \alpha \forall x \in R$.

The *support* of a real valued function f is given by

$$Supp(f) = \{x \mid f(x) \neq 0\} \ .$$

Let $O(x) = \sum_{i=1}^{N} \chi_j(x)$ be the *overlap function* of the N abstract sensors. For each $x \in \Re$, $O(x)$ gives the number of sensor intervals in which x lies, that is, the number of intervals overlapping at x.

An N-clique is a group of N intervals having a common intersection.

From the definitions, it follows that the integer $\|\chi_i O(x)\|$ gives the maximum number of intervals in which any $x \in f_i$.

Note that $S(x)$ is the support of the characteristic function of all those points that belong to the intersection of $(N - F)$ or more intervals. The correct physical value belongs to Supp($S(x)$), that is, to one of the intervals constituting it. Algorithm M1 returns the smallest connected interval containing Supp($S(x)$) for the integrated output.

More precisely, the output interval estimate I_p is given by

$$I_p = [\min\{x|S(x) = 1\}, \max\{x|S(x) = 1\}]$$

Thus, algorithm M1 gives a connected interval within which the actual physical values lie.

Note the result of M1 includes points that do not belong to the intervals constituting Supp($S(x)$). Furthermore, if the intervals constituting Supp($S(x)$) are widely scattered over the union of I_is, then I_i suffers from inaccuracy since it tends to be very broad.

8.1.2   Characterizing Algorithm PIKM

Algorithm PIKM distinguishes between wild faults and tame faults. The rationale is as follows. A sensor fails wildly if there is no correlation between the actual physical value being measured and the interval estimate of the faulty sensor. If a sensor fails tamely, the faulty sensor's interval estimate does not contain the actual physical value but lies significantly close to the value in a certain sense. Since we do not know the actual physical value, we cannot directly detect the tameness of a fault. However, tamely faulty sensor estimates tend to overlap with correct sensor estimates because of their proximity to the actual physical value.

Correct sensors tend to have a relatively larger number of overlapping intervals as compared to undetected faulty sensors participating in the $(N - F)$-intersections, since tamely faulty sensors overlap with correct sensor estimates. Thus, the number of sensor estimates overlapping with a given sensor estimate is a good index of its correctness. This observation helps us to narrow the output interval estimate, I_p.

Let

$$Supp(S(x)) = \bigcup_{i=1}^{k} L_i$$

where $L_i = [\alpha_i, \beta_i]$ with $\beta_i < \alpha_{i+1}$ $(1 \leq i \leq k - 1)$.

Evaluate L_i's and attach weights to them. Choose those L_i's with maximum weight to be the intervals that have a high likelihood of containing the correct physical value. Enclose these L_i's of maximum weight by the smallest possible interval and take it to be the output estimate.

Let $\chi_L i(x)$ be the characteristic function of L_i. Then the popularity of the jth sensor is the number

$$P_j = \sum_{k=1}^{n} \|c_k \ c_j\| - 1$$

P_j gives the number of sensor intervals overlapping with the jth sensor interval.

The sum of the popularities of all sensors involved in the formation of L_i is called the reliability r_i of L_i. The *width* of the output interval estimate can be narrowed by using *popularity* and *reliability*.

Consider the set function

$$W(L_i) = \sum_{j=1}^{n} \|\chi_{L_i} \chi_j\| P_j$$

$(1 \leq i \leq k)$ defined on each L_i.

$W(L_i)$ gives the sum of the number of intervals overlapping with each sensor estimate in the $(N - F)$-or-more clique L_i. that is,

$$r_i = W(L_i) \forall 1 \leq i \leq k .$$

Let

$$r = \max\{r_i | 1 \leq i \leq k\}$$
$$m = \min\{i | r_i = r\}$$

and

$$M = \max\{i | r_i = r\}$$

Consider the interval $[\alpha_m, \beta_M]$. We take $I_p^* = [\alpha_m, \beta_M]$ as the integrated output estimate.

Algorithm PIKM is no worse than M1 for a one-dimensional sensor integration problem. I_p = output estimate given by M1. I_p^* = output estimate given by PIKM. Thus,

$$\beta_M - \alpha_m = |I_p^*| \leq |I_p|$$

where $|I|$ is the *width* of the interval I.

In the PIKM's model, there is in general a way of narrowing the output estimate I_p to I_p^*. The improvement depends on two main assumptions: that the number of sensors is large and that most faults are tame. If one assumption does not hold, it is possible (in the worst case) that $|I_p^*| = |I_p|$. If the number of wildly faulty sensors as many as the tamely faulty ones and if they happen to cluster somewhere else on I_p, it is possible that $|I_p^*| = |I_p|$. Thus, the worst case for I_p^* is I_p. The chances that wildly faulty sensors mimic the clustering behavior of tamely faulty sensors are remote. Also, if the number of sensors is very small, it is possible that $|I_p^*| = |I_p|$.

Example 1

In Fig. 8.1, the numbers near each interval estimate give the number of intervals overlapping with it. Here $N = 13$ and $F = 10$. The thick lines in the figure are the intervals L_i, with the numbers on them indicating their reliabilities. We may pick the interval with higher reliability. Alternatively, may define a range for reliabilities and pick intervals that fall in these limits. The thick line at the bottom indicates the output interval estimate for this case in Marzullo's model. We can see that algorithm PIKM provides better performance that algorithm J1.

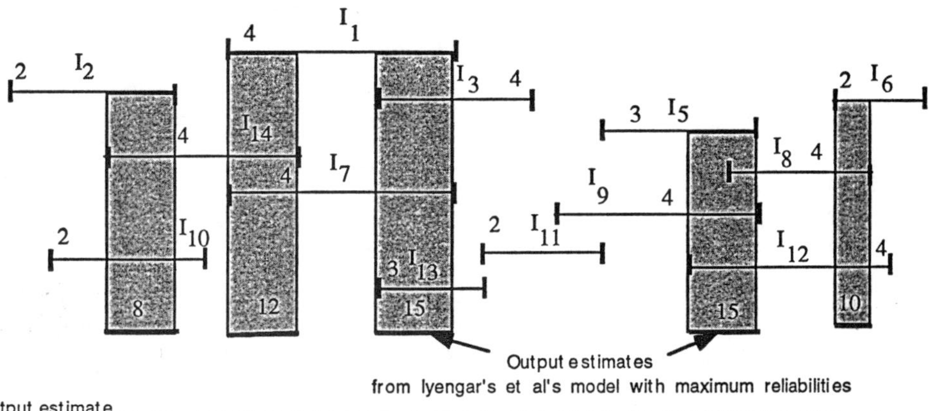

Popularities of intervals.

Interval	I_1	I_2	I_3	I_4	I_5	I_6	I_7	I_8	I_9	I_{10}	I_{11}	I_{12}	I_{13}
Popularity	4	2	4	4	3	2	4	4	4	2	2	4	3

Figure 8.1 Integration of intervals ($N = 13$, $F = 10$).

Example 2

Let the three input sensor estimates be $I_1 = [l_1, u_1]$, $I_2 = [l_2, u_2]$, $I_3 = [l_3, u_3]$, and $l_1 < l_2 < u_1 < l_3 < u_2 < u_3$. The estimate given by algorithm M1 $= I_p = [l_2, u_2]$. By the *polling* technique, there are two intervals $L_1 = [l_2, u_1]$ and $L_2 = [l_3, u_2]$, but they both have the *same* reliability $= 2$. So, $I_p^* = [l_2, u_2]$. Hence $I_p^* = I_p$ here.

Let $l_1 = 2.4$, $u_1 = 3.2$, $l_2 = 2.9$, $u_2 = 4.0$, $l_3 = 3.6$, and $u_3 = 5.0$. With a large number of sensors, we can see that the clustering of the tamely faulty sensors gives a clue to the whereabouts of the correct value. Correspondingly, the width of the output intervals is reduced greatly, as compared to the output interval estimate given by M1.

8.2 Computational Framework for DSNs

A computational framework for distributed sensor networks should be general enough
to incorporate existing distributed sensor networks. It should also allow us to characterize *distributed sensor integration*. In particular, it should address the following:

- Fault-tolerant fusion of information from multiple sensors

- Mapping and modeling the environment space

- Task-level complexity issues of the computational model

- Formalization of a wide class of distributed sensor networks

A formal description for an abstract DSN should include the following:

- Be a very theoretical framework for characterizing a general DSN

- Address the general computational features and problem of a DSN

- Symbolically describe the features of a large class of DSNs

- Not explicitly deal with implementation of the DSNs

Distributed sensor networks differ in implementation details. This is partly due
to the nature of the phenomena being sensed and the sensors. We are interested in
a model that can describe formally an abstract distributed sensor network with the
ability to incorporate specific implementations of the abstract structure. A key idea
is to look on sensor information as geometric point sets rather than as distributions.

An abstract distributed sensor network is a 4-tuple

$$(X, \{E_i\}_{i=1}^m, \{S_i\}_{i=1}^m, P)$$

where:

1. X is the *environment space*. It is the space under observation by the sensors of
the distributed sensor network. Clusters of sensors are allocated to various regions
of X, and these clusters pick up data from their respective regions and transmit
them to the processors to which they are connected. The collection of regions to
which the various sensor clusters are allocated constitute the environment space X.

2. P is the *parameter space*. It is the space of all possible values of the parameter
being measured. If the parameter being measured is a k-dimensional vector, then
P is the *Euclidean space* of dimension k. The value of each sensor is represented by
a subset of P.

3. $\{E_i\}_{i=1}^m$ is a *chart* on X. It is a collection of subsets of the environment space
X such that their union is X. (see Fig. 8.2). Each subset E_i is allocated a cluster
of sensors that obtains data from E_i and reports to a processor.

4. $\{S_i\}_{i=1}^m$ is a collection of m sensor clusters, each S_i being assigned to an
E_i. Each S_i is a cluster of n_i abstract sensors $S_i = \{\sigma_{i,j}^t\}_{j=1}^i$, where each abstract

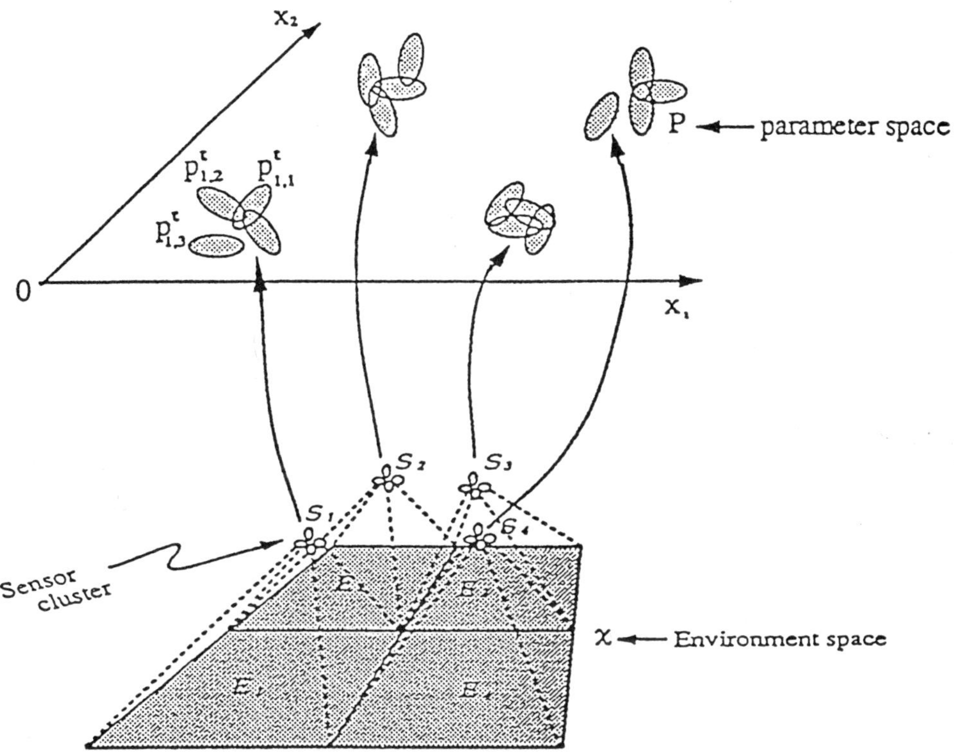

Figure 8.2 Schematic representation of an abstract DSN.

sensor $\sigma^t_{i,j}$ is a time-varying map that maps the set E_i onto a subset $p^t_{i,j}$ of $X \times P$. The values of the subsets $p^t_{i,j}$ are described later. Thus, at any instant of time t we have the collection of subsets $\{p^t_{i,j}\}^{n_i}_{j=1}$ of $X \times P$ as the abstract sensor estimates of the sensors $\{\sigma^t_{i,j}\}^i_{j=1}$ of the parameter values observed in E_i at time t.

The sensor $\sigma^t_{i,j}$ senses data about a parameter at each point $x \in E_i$. Since a sensor in general has some uncertainty in its reading, it is realistic to expect it to give a range of values for the parameters' values at the point x instead of a single value. Thus we assume that for each $x \in E_i$, $\sigma^t_{i,j}$ gives a connected subset of values representing the value of the parameter at x. If the actual value of the parameter at x belongs to this set of values, then $\sigma^t_{i,j}$ is correct at the point x, else it is faulty.

Thus $\sigma_{i,j}^t$ maps $x \in E_i$ onto the subset $p_{i,j}^t$ of $X \times P$ given by $p_{i,j}^t(x) = \{(x,z) \mid z$ is in the range of values given out by $\sigma_{i,j}^t$ at $x\}$.

We assume $p_{i,j}^t(x)$ is connected for each $x \in E_i$. We designate $p_{i,j}^t = \bigcup_{x \in E_i} p_{i,j}^t(x)$. If the parameter manifests itself only at finitely many discrete points $x_1, x_2, \ldots, x_k \in E_i$ then $p_{i,j}^t = p_{i,j}^t(x_1) \cup p_{i,j}^t(x_2) \cup \ldots p_{i,j}^t(x_k)$. If the parameter manifests itself on finitely many connected subsets of E_i, then $p_{i,j}^t$ itself is a union of finitely many connected subsets of $X \times P$ or, more specifically, of $E_i \times P$.

Thus, in any case, we see that $p_{i,j}^t$ is a set of finitely many disjoint connected subsets of $X \times P$. The dimension of the sets $p_{i,j}^t(x)$ is in general the same as the dimension of the parameter measured by the sensor $\sigma_{i,j}^t$.

Each sensor $\sigma_{i,j}^t$ in the cluster S_i collects data from the same region in the environment space X, that is, E_i. Under ideal fault-free conditions, all the $\sigma_{i,j}^t$ in S_i have identical outputs.

The set $p_{i,j}^t$ represents readings of the sensor j in the cluster i. Here, it is assumed that it reads the parameter values at all points of the one-dimensional interval set E_i. $\sigma_{i,j}^t$ returns an interval of values $(p_{i,j}^t(x))$ as the abstract sensor reading at the point x. Here the parameter is assumed to be one-dimensional.

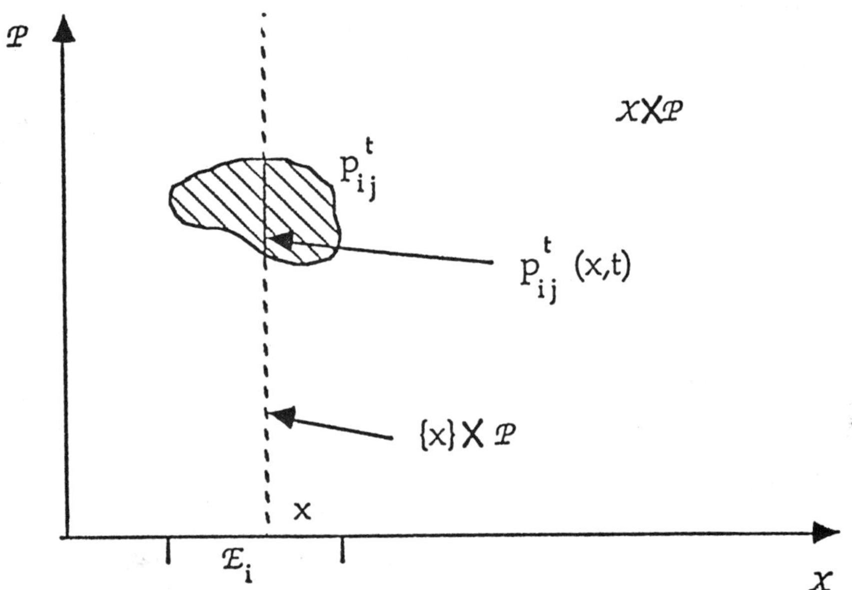

Figure 8.3 Subsets $p_{i,j}^t$ corresponding to sensor $\sigma_{i,j}^t$.

The schematic anatomy of the set $p_{i,j}^t$ is shown in Fig 8.3.

8.2.1   Boiler

Example 1

Consider a boiler with clusters of temperature probes placed at various locations in it. The environment space consists of all the points where the various clusters of temperature probes are placed. Since temperature is a scalar, the parameter space P is one-dimensional Euclidean space. The sets E_i are the points at which the clusters of temperature probes are placed. The abstract sensor $\sigma_{i,j}^t$ corresponding to a probe is a map that associates with the point x monitored by the cluster of probes to which the probe belongs a one-dimensional line segment (or a 1-interval)

$$p_{i,j}^t(x) = p_{i,j}^t = \{(x,y)|y \in P, a_j \le y \le b_j\}$$

where a_j and b_j are the minimum and maximum values registered by the sensor $\sigma_{i,j}^t$ (see Fig. 8.4).

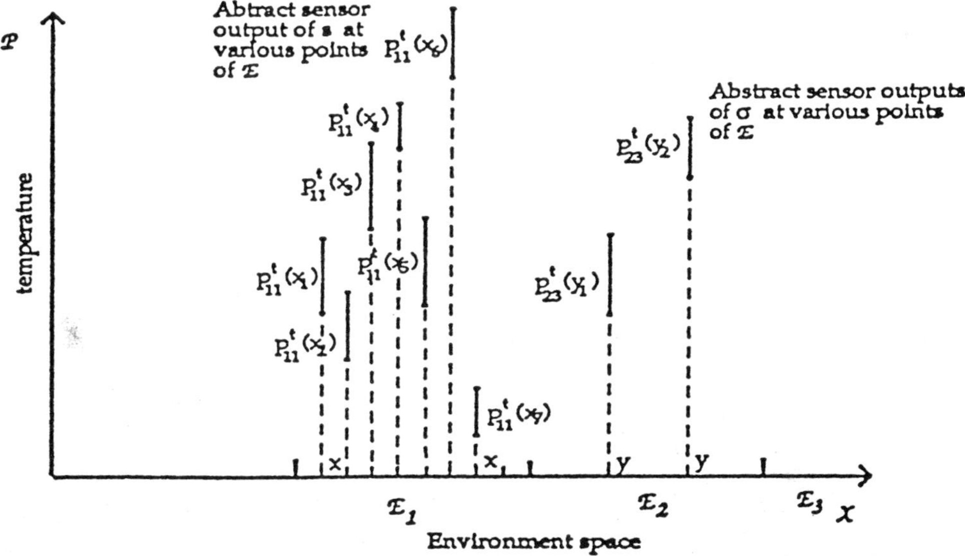

Figure 8.4 Schematic representation of the four-dimensional space $X \times P$.

8.2.2 Aerial Mapping

Example 2

Consider aerial mapping of a terrain. The terrain may be divided into regions and clusters of sensors allocated to each region. Each sensor measures the altitudes of various points in the region to which it is assigned. It is modeled by an abstract sensor $\sigma_{i,j}^{t}$ that returns a one-dimensional range of values for each point in the region. The regions are the sets E_i. The parameter space P is a one-dimensional space since altitude is a scalar. Since the sensors $\sigma_{i,j}^{t}$ register the altitude of each point in E_i, the subset $p_{i,j}^{t}$ is a three-dimensional connected subset of $X \times P$, which is 3-dimension (see Fig. 8.5).

8.2.3 Tracking

Example 3

Consider the problem of tracking and reading the velocities of flying aircraft in a given air space. Suppose the sensors employed are capable of detecting and reading the velocities of objects in a region of certain size. We might consider a decomposition of the air space under observation into regions of appropriate sizes to fit the scope of the sensors. Each sensor is modeled by an abstract sensor $\sigma_{i,j}^{t}$ that associates with each point x of its region a connected three-dimensional subset as the range of values for the velocity of the object at the position x. The parameter space P is three-dimensional Euclidean space since velocity is a three-dimensional vector. The environment space X is the space of all observable positions and is thus three-dimensional. For each $x \in E_i$, $p_{i,j}^{t}(x)$ is a connected three-dimensional subset of $X \times P$ and $p_{i,j}^{t}$ is a collection of disjoint connected three-dimensional subsets of $X \times P$ (see Fig. 8.6).

8.3 Model of Sensors

An abstract sensor $\sum_{i,j}^{t} \in S_i$ is correct at $x \in E_i$ at time t if $(x, a) \in p_{i,j}^{t}(x)$, where a is the actual value of the parameter being measured at x at time t $(a \in P)$; else it is *faulty*.

Consider the *characteristic function* of the set $p_{i,j}^{t}(x)$:

$$X_{x,t}^{i,j}(y) = \begin{cases} 1 & \text{if } y \in p_{i,j}^{t}(x) \\ 0 & \text{otherwise} \end{cases}$$

Note that the characteristic function of the set

$$p_{i,j}^{t} = \bigcup_{x \in E_i} p_{i,j}^{t}(x)$$

is given by

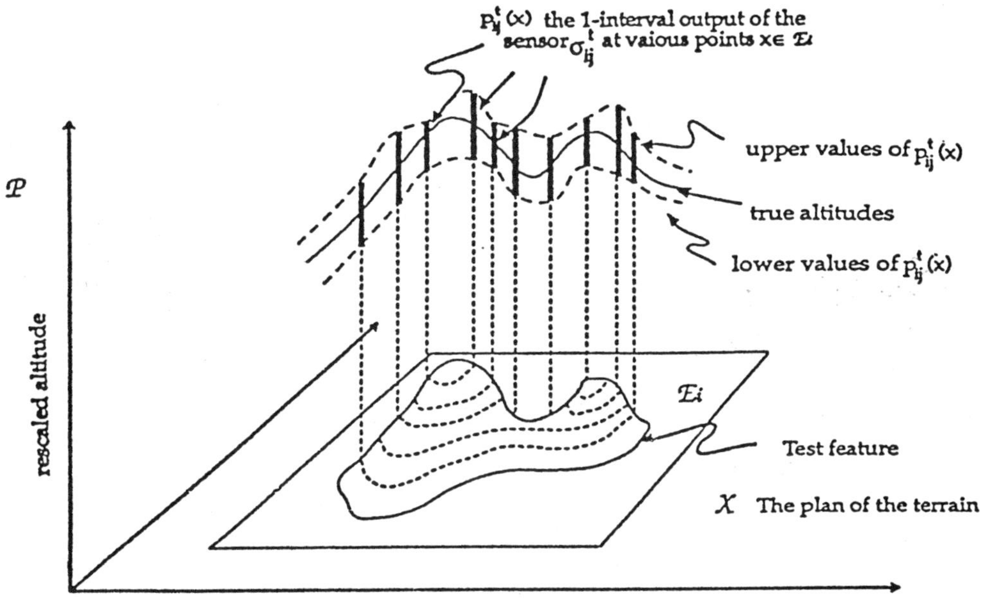

Figure 8.5 Actual representation of $p_{i,j}^t(x)$ for various $x \in E_i$ in Example 2.

$$X_t^{i,j} = \sum_{x \in E_i} X_{x,t}^{i,j}(x), (x \neq y \Rightarrow p_{i,j}^t(x) \cap p_{i,j}^t(x) = \emptyset)$$

If at most f_x of the n_i sensors in the cluster S_i are known to be faulty at x, then following the method of integration discussed in Chapter 2, the correct value of the parameter being measured by the sensor cluster s_i at $x \in E_i$ and at time t is contained in the subset corresponding to the characteristic function $X_{x,t}^i(y)$ given by

$$X_{x,t}^i(y) = X[n_i - f_x, \infty)(O_{x,t}^i(y))$$

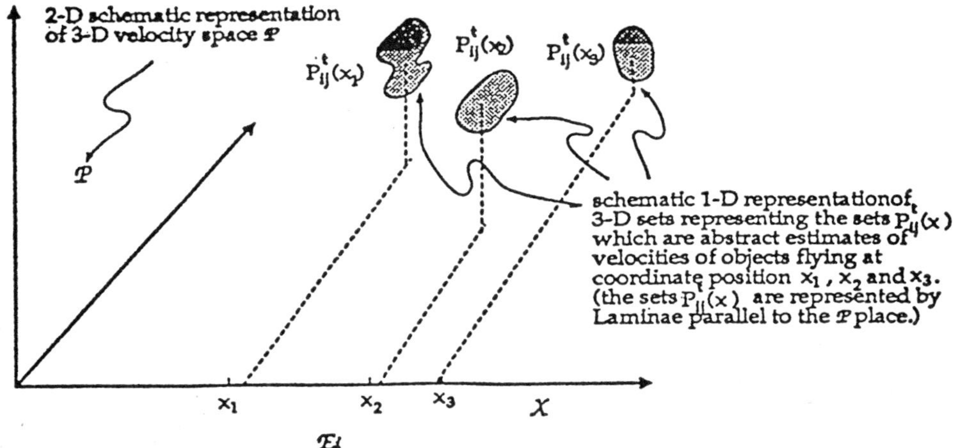

Figure 8.6 Schematic one-dimensional representation of $X \times P$ in Example 3.

where

$$X[a, \infty)(x) = \begin{cases} 1 & \text{if } x \in [a, \infty) \\ 0 & \text{otherwise} \end{cases}$$

and

$$O_{x,t}^i(y) = \sum_{j=1}^{n_i} X_{x,t}^{i,j}(y)$$

is the overlap function for the cluster S_i at the point $x \in E_i$ and at time t. That is, $O_{x,t}^i(y)$ gives the number of sensors $\sum_{i,j}^t$ whose estimates $p_{i,j}^t(x)$ at the point $x \in E_i$ and at the time t contain the point $(x, y) \in \{x\} \times P, y \in P$. Consequently, $X_{x,t}^i(y)$ is the characteristic function of the set of all those points of $\{x\} \times P$ that lie in the intersection of at least $n_i - f_x$ intersections of the sets $p_{i,j}^t(x)$. The correct value of the parameter being measured at $x \in E_i$ at time t must lie in the subset $S_i(x, t) = \{y \in P | X_{x,t}^i(y) = 1\}$, since the assumption that at most f_x sensors are faulty at x implies at least $n_i - f_x$ sensors are correct and hence must overlap, since their estimates $p_{i,j}^t(x)$ must all contain the point (x, a) where a is the actual value of the parameter being measured.

As there is no reason to assume that f_x is a fixed or bounded quantity, we may as well treat it as a parameter that we may change at will to obtain larger or smaller intersections as desired.

For instance, we may look on f_x as

$$\min \left\{ f \,\middle|\, X^i_{x,t}(y) = X_{[n_i-f,\infty)}(O^i_{x,t}(y)) \neq 0 \right\}$$

and thus obtain those intersections that are a result of the largest number of sets intersecting to give a nonempty set.

Faulty sensors could have *random* or *wild* faults, in which event there is little correlation between the sensor's estimate and the correct physical value of the parameter being measured. On the other hand, the fault may be *tame*, in which case the faulty sensor's estimate does not contain the actual physical value of the parameter being measured, but lies sufficiently close to it. It is reasonable to assume that most sensors fail tamely due to perturbations encountered in the physical world, in which case faulty sensors tend to cluster in the neighborhood of the correct value of the parameter. We may make use of this clustering to predict with reliability the subset with the highest chance of containing the actual value of the parameter among those subsets that belong to $\sum^i(x,t)$.

To do this, we need to make the notion of a tame fault more rigorous. The following definition is derived from the above characteristics of tame faults.

A *faulty* sensor is *tamely faulty* at x at time t if it intersects with a sensor that is correct at x at time t. It is unlikely that sensors with wild or random faults cluster since by their very nature they are uncorrelated and hence distributed more or less evenly.

8.3.1 Characterizing Polling Technique

If most sensor faults are tame, a polling technique can be used to reduce the measure of the subset containing the correct physical value with high reliability. Let $L^i_1(x,t), \ldots, L^i_k(x,t)$ be the disjoint maximal connected subsets of P whose union is the subset $\sum^i(x,t)$ containing the correct physical value of the parameter measured at x at time t $(x \in E_i)$.

Define the *popularity* of the output of the kth sensor $(1 \le k \le n_i)$ in the cluster S_i to be the nonnegative integer $\pi^{ik}(x,t)$ given by

$$\pi^{ik}(x,t) = \left(\sum_{j=1}^{n_i} \| X^{ij}_{x,t} X^{ik}_{x,t} \| \right) - 1$$

where $\|f\|$ is the maximum value of the function f. $pi^{ik}(x,t)$ gives the number of sensor outputs having nonempty intersection with the output of the sensor $\sigma^t_{i,k}$ at x and time t.

If $\Lambda^{ij}_{x,t}$ is the characteristic function of the set $\Lambda^i_j(x,t)$ $(1 \le j \le k_i)$ then define the reliability $R^{ih}(x,t)$ of the set $\Lambda^i_j(x,t)$ to be the nonnegative integer given by

$$R^{ik}(x,t) = \sum_{k=1}^{n_i} \|\Lambda^{ij}_{x,t} X^{ik}_{x,t}\| \pi^{ik}(x,t)$$

$R^{ik}(x,t)$ is the sum of the popularities of all those sensors' outputs that contain the set $\Lambda^i_j(x,t)$.

As the reliability of the connected set $\Lambda^i_j(x,t)$ becomes greater, there is more clustering of sensors about $\Lambda^i_j(x,t)$, and hence the likelihood of $\Lambda^i_j(x,t)$ containing the correct value of the parameter increases. Thus, $R^{ij}(x,t)$ is a good measure of the connected subsets $\Lambda^i_j(x,t)$ of $\sum^i(x,t)$ containing the correct value of the parameter. [It is obvious that one and only one of these maximal connected subsets of $\sum^i(x,t)$ contains the correct parameter value]. The maximal connected subsets with the highest reliability may be chosen to represent the estimate of the sensor cluster S_i.

This analysis greatly reduces the measure of the output subset of S_i and is an efficient fault-tolerant integration technique for the case when most sensor faults are tame. Although our techniques discuss one-dimensional interval estimates, they hold for a much more general class of sensor outputs.

If the sensors provide additional information on their outputs (e.g., a probability distribution or a weight function on their output values), we may replace the characteristic function of the output sets by these functions and look for an appropriate integration method that combines these functions (in such a way that the correctness of the output is not sensitive to a small number of faults in the sensor cluster) to give a function whose domain contains the correct value of the parameter measured by the cluster, and the correct value lies in a high-weighted or high-probability region of this new function/distribution. Depending on the kind of extra information provided by the sensors and the functions involved, the method of integration for these functions varies.

All the above analysis is local to each subset E_i of the chart on the environment space X. The integration of the sensor outputs in each cluster is done by the processor allocated to that cluster. Thus the parameter measured in each subset of the chart on X is measured and estimated by the sensor cluster-processor unit of that subset. The subsets of the chart on X form a tiling (tessellation) of the environment space X. If the sensors are probes capable of measuring a parameter's value only at a given point, then X will be the collection of all these points.

If the sensors are capable of sampling a region, then the space X is the union of all these regions. The sampling of the sensor outputs is done either periodically in time or randomly. But they are all sampled at the same instant of time, for otherwise integration of the sensor values would not make sense. We may then proceed to get an overall picture of the behavior of the parameter over X by studying the outputs of all the sensor clusters. This is done by appropriately juxtaposing the sensor values according to the actual geometric layout of the chart and obtaining the profile of the parameter over the space over which the sensor clusters are distributed.

It is also possible that the processors allocated to the sensor clusters may have to communicate among themselves for efficient overall performance. For example, in the case when the sensors are tracking a moving object and the object moves from one set in the chart to another, the sensor cluster currently observing the object activates the neighboring sensor cluster, which prepares to track the object from an indicated position on the boundary of the chart set it is allocated to. Here, we may take advantage of this kind of interprocessor communication to check for faults by comparing common values. It is therefore often fruitful to overlap the sets in the chart to obtain extra redundancy and greater fault tolerance.

8.3.2   Global Behavior of a Parameter

This involves the interaction of the processors allocated to the subsets of the chart and comparison of common information for smooth patching of local information to obtain a global picture of the phenomenon being observed over X.

Each of the subsets of the subsets

$$E_i(1 \leq i \leq m, \bigcup_{i=1}^{m} E_i = X)$$

covers a region of the environment space X and is equipped with a sensor cluster

$$S_i = \{\sigma_{i,j}^t\}_{j=1}^{n_i}$$

where each sensor $\sigma_{i,j}^t$ monitors all of E_i and sends data to a common processor allocated to E_i. Each E_i is equipped with its own coordinate system, and all the sensors in the cluster S_i measure with reference to this coordinate system. Thus, all points x in E_i have local coordinates, and if two of the regions E_i overlap then the coordinates of a point in the intersection will be different in the different regions containing it. The relative arrangement of the subsets E_i of the chart on X depends on the specific needs of a distributed sensor network. However, it is desirable to have them overlapping since this helps in patching up the local scenes to form a global picture, as well as in increasing fault tolerance and aiding fault detection in sensors by comparison of data at common points. The union of the outputs of the processors over the sets E_i gives the global profile of the parameter over X.

For example, in the case when a distributed sensor network is mapping a terrain by measuring altitudes at various points, each processor yields an output that is an estimate of the altitude of each point in E_i. The patching together of the processor information yields a global profile of the altitude of the terrain X at each point.

The sets E_i may be overlapped to corroborate the relative locations of these sets and enhance fault tolerance and fault-detection by pitting one processor's output against the other and cross-checking to find out if any cluster has failed (massive failure). We may also conceive of other specification and need-based variations of the relative arrangements of the E_i, for instance, when these sets are themselves in motion with relation to one another.

In Fig. 8.7, regions E_i, E_j, and E_k are under observation by sensor clusters $\{\sigma_{i,j}^t\}_j$, $\{\sigma_{j,l}^t\}_l$ and, $\{\sigma_{k,m}^t\}_m$. p_i, p_j, and p_k are processors associated with the clusters i, j and k. They are linked to one another to communicate and compare information common to them on the sets $E_i \cap E_j$ and $E_j \cap E_k$. This comparison leads to proper juxtaposition of outputs, as well as fault tolerance and fault detection at the processor level.

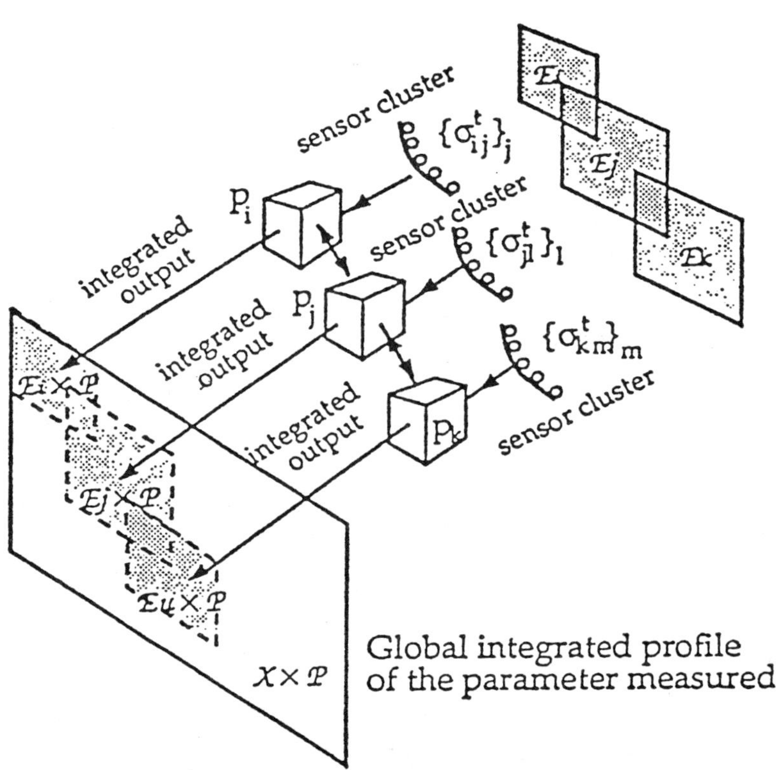

Figure 8.7 Schematic representation of information patching to obtain a global picture.

Let E_i and E_j be two sets in the chart on X that overlap and a point x in the intersection has local coordinates $(x_1^i, \ldots, x_k^i)$ and $(x_1^j, \ldots, x_k^j)$ in the sets E_i and E_j, respectively. We can obtain the local coordinates of x w.r.t E_i to the local coordinates of x w.r.t E_j using the coordinate transformation T_{ij} defined on $E_i \cap E_j$.

$$T_{ij}(x_1^i, \ldots, x_k^i) = (x_1^j, \ldots, x_k^j)$$

Clearly, transformations T_{ij} and T_{ji} are inverses of each other on $E_i \cap E_j$.

For instance, the E_i may be k-dimensional hypercuboids of identical dimensions overlapping in some manner and with Cartesian local coordinates. Then the T_{ij} are affine linear transformation involving a rotation and a translation (see Fig. 8.8).

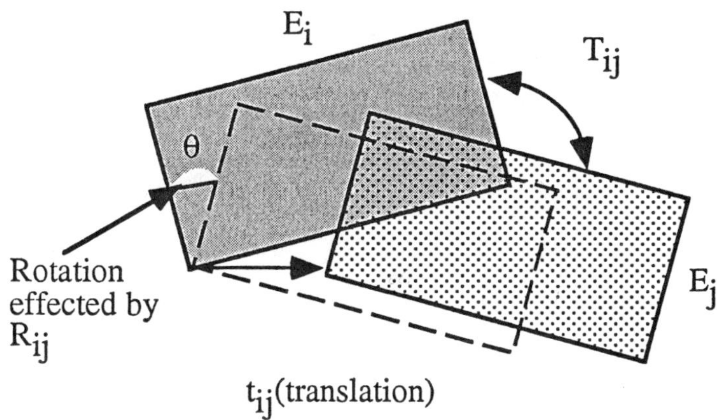

Figure 8.8 Linear transformation involving a rotation and a translation for T_{ij}.

There exists a $k \times k$ matrix R_{ij} and a k-dimensional column vector t_{ij}, such that

$$(x)_j = R_{ij} \bullet (x)_i + t_{ij}$$

where $(x)_j$ and $(x)_i$ are column vector representations of the local coordinates of x in E_j and E_i, respectively. That is, i.e.

$$T_{ij}(x)_i = R_{ij} \bullet (x)_i + t_{ij}$$

8.3.3   Linear Transformations

Example 1

of a two-dimensional environment space. Let p be a point lying in the intersection of r_1 and r_2 (Fig. 8.9). Let (x_1, y_1) be the coordinate of p w.r.t. the coordinate system of E_1. Let (x_2, y_2) be the coordinate of p w.r.t. the coordinate system of E_2. If $v = (x_0, y_0)$ is the displacement vector of the origin of the coordinate system of r_1 with respect to the coordinate system of r_2, then we have

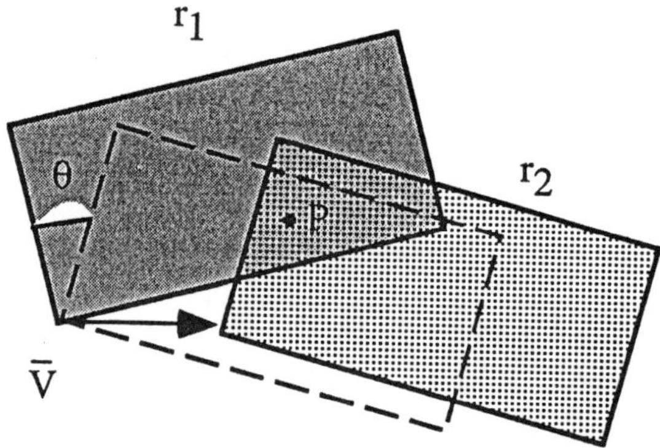

Figure 8.9 Example of linear transformation.

Let r_1 and r_2 be the rectangles representing the sets E_1 and E_2

$$\begin{pmatrix} x_1 \\ y_1 \end{pmatrix} = \begin{pmatrix} \cos\theta & -\sin\theta \\ \sin\theta & \cos\theta \end{pmatrix} \begin{pmatrix} x_2 - x_0 \\ y_2 - y_0 \end{pmatrix}$$

$$X_1 = R_\theta X_2 + t_2$$

where R_θ is a rotation by θ and t_2 is a translation.

8.3.4   Transition Functions

Suppose the E_i's are not rigidly fixed but move in time and space, as is the case with tracking sensor systems. Then the transformations T_{ij} are time varying. In the case when there is only one chart set and it is in motion, the time varying transformation that allows for the movement of the sensor system corresponds to an inertial navigation system. A chart equipped with local coordinates is called a coordinate chart on X, and the transformations that effect local coordinate changes are called transition functions. The transition functions are known beforehand if the chart sets are fixed. If the chart sets move in a predictable or prescribed way then again the time-varying transition function can be computed.

Figure 8.10 gives a schematic representation of overlapping chart sets of an m-dimensional environment space. Figure 8.11 describes the transition functions associated with the chart.

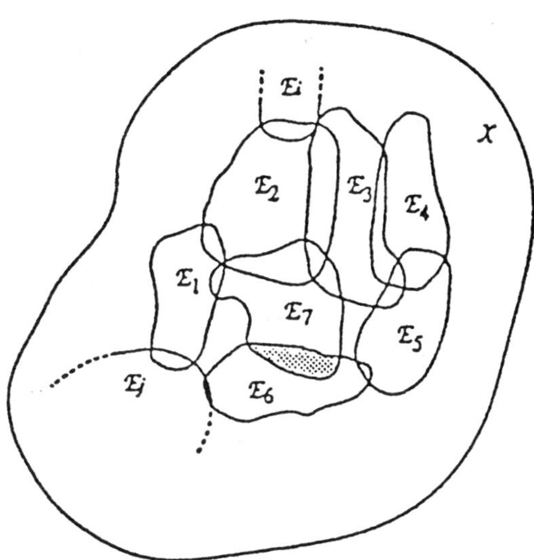

Figure 8.10 A general coordinate chart $\{E_i\}_{i=1}^{n}$ on an environment space X.

The general transformation scheme between two sets E_i and E_j is shown. Let $p \in E_i \cap E_j$; then

$$p = (x_1^i, \ldots, x_m^i) \in E_i$$

$$p = (x_1^j, \ldots, x_m^j) \in E_j$$

ϕ_i: the transition function from E_i to E_j defined on $E_i \cap E_j$
ϕ_j: the transition function from E_j to E_i defined on $E_i \cap E_j$
$\phi_i \bullet \phi_j = $ identity on $E_i \cap E_j$ w.r.t E_j
$\phi_j \bullet \phi_i = $ identity on $E_i \cap E_j$ w.r.t E_i
The coordinates of p in E_j are obtained from the coordinates of p in E_i using ϕ_i to transform p's coordinates from E_i to E_j:

$$\phi_i(x_1^i, \ldots, x_m^i) = (x_1^j, \ldots, x_m^j)$$

Similarly,

$$\phi_j(x_1^j, \ldots, x_m^j) = (x_1^i, \ldots, x_m^i)$$

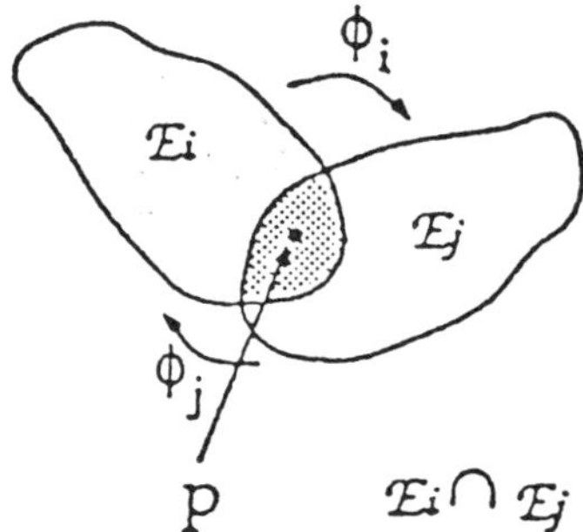

Figure 8.11 Transition functions ϕ_i and ϕ_j defined on the intersection $E_i \cap E_j$.

Figures. 8.12 and 8.13 describe two simple examples of overlapping chart sets on two-dimensional environment spaces. In Fig. 8.12, a two-dimensional rectangular region X is tiled by partially overlapping rectangular chart sets. Two neighboring rectangular sets overlap on a strip. The parameter values on these strips are common to both adjacent sensor clusters and may be compared for effective patching together of local information to obtain a global picture. In this case, the transition functions are horizontal or vertical translations by a fixed distance in each direction.

In Fig. 8.13, a two-dimensional circular region X is tiled by partially overlapping sectors. These sectors correspond to the sets E_i. Two neighboring sectors overlap on a narrow subsector. The parameter values on these strips are common to both adjacent sensor clusters and may be compared for effective patching together of local information to obtain a global picture. In this case, the transition functions are clockwise or counterclockwise rotations through a fixed angle. Further analysis in this direction can be done in specific distributed sensor network situations, but the general interprocessor interaction and communication must involve the use of a transition function for meaningful comparison of data. The coordinate changes are effected by local processors, and only those processors that are connected have transition functions defined. In the case of moving chart sets, it is possible that two sets whose processors are not connected directly may overlap for some time. It is possible to perform local coordinate changes by routing data through other intermediate processors since the transition functions are transitive and hence may be composed to obtain transformations between unconnected processors. It is to be noted that the overlapping of chart sets not only helps in comparing sensor values to detect sensor faults, increases fault tolerance at the sensor level, and patches up local information to obtain global information, but it can also be utilized to

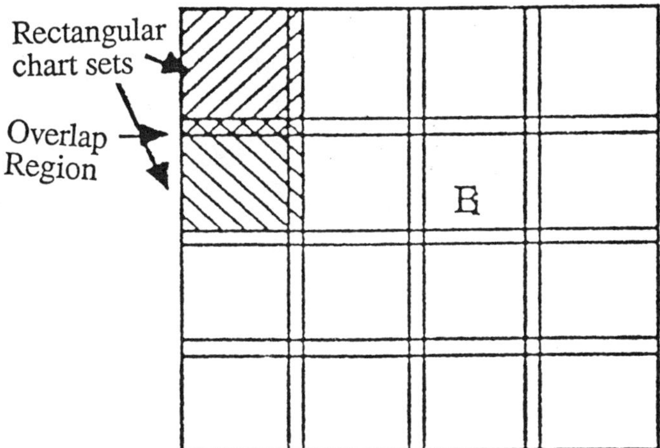

Figure 8.12 A two-dimensional rectangular region X tiled by partially overlapping rectangular chart sets. The transition functions here are just horizontal or vertical translation.

compare processor values to detect processor faults and incorporate fault tolerance at the processor level.

8.4 Chapter Summary

In this chapter we showed how sensor integration algorithms can be formalized. We first formalized the algorithm M1 (by Marzullo) and then the algorithm PIKM (by Prasad, Iyengar, Kashyap, and Madan). We then formalized the three-interval sensor integration algorithm used in the flat tree network and the multilevel inary de Bruijn network. Finally, we gave a computational framework for describing abstract distributed sensor networks.

8.5 References and Further Reading

This chapter is adapted from [60] and [92].

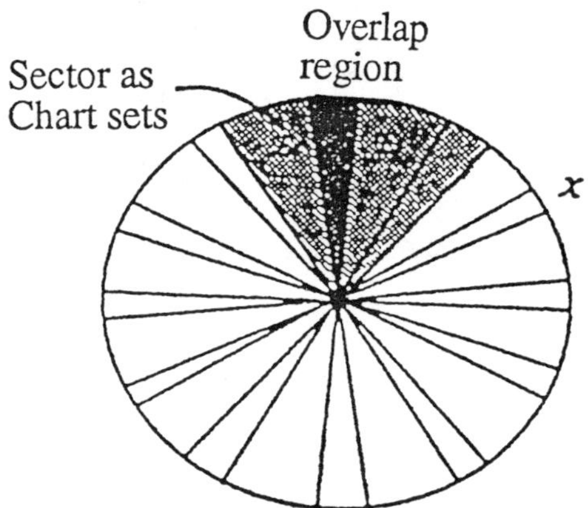

Figure 8.13 A two-dimensional circular region X tiled by partially overlapping sectors. The transition functions here are clockwise and counterclockwise rotations.

Problem Set 8

8-1. Name three regular tilings.

8-2. How can we extend the concept of a correct sensor and a faulty sensor to the time-varying case?

Chapter 9

Distributed Detection Problems

We have discussed several aspects of sensor integration. In this chapter, we will take a look at a class of problems that involve distributed detectors.

A sensor is a device capable of sensing a physical value. It is called a detector if the sensor is used to detect an object or an event. Microwave radars, laser radars, and infrared sensors are examples of detectors used in an air surveillance system. They are used to detect and keep track of flying objects (e.g., enemy planes) in the air space.

An autonomous mobile robot is equipped with a vast array of sensors: vision sensors, range finders, and touch sensors. Objects such as chairs and dogs might lie in the path of a mobile robot. Such objects, considered obstacles by the robot, must be detected and avoided.

A detection system employs a finite number of sensors. A sensor may or may not be able to detect all objects that are present in its work space. Likewise, an object may or may not be detected by all sensors in the detection system.

Depending on the restrictions placed on the nature of the sensors and the objects, we obtain a class of *detection problems*. First, we may give a list of sensors for a detection system and, for each sensor, we specify the set of objects it can detect. Second, we may give a list of objects that can be present in a detection system and, for each object, we specify the set of sensors that can detect the object.

Generally, there is a many-to-many relationship between the set of sensors and the set of objects. A sensor can detect one or more objects; and an object can be detected by one or more sensors. However, due to technological or other constraints, a sensor might not be able to detect all objects.

The detection may take place in varying degrees. A binary detector returns 1 (or true) if it detects an object and 0 (or false) otherwise.

A detector may also be required to identify the object being detected. A detector might use hypotheses, confidence values, and belief intervals to come up with a local decision. A reliable detector should come up with the correct decision; however, there is a slight chance that it might miss an object or that it might raise a false alarm.

The use of several detectors requires a global decision to be made using the local decisions. This is called *decision fusion*. Three commonly used models of decision fusion are the (1) serial model, (2) parallel model, and (3) multiple level parallel decision fusion (MLPDF) model [49, 48].

In this chapter, we are concerned with simple versions of the *distributed detection problem*, that is, involving multiple sensors and multiple objects.

Tsitsiklis and Athenas [78] discuss the complexity of decentralized decision making and detection problems. Rao [62] discusses the computational complexity issues in the synthesis of simple distributed detection networks. By imposing several restrictions on the distributed detection problem, we should obtain problems that are conceptually simpler. Although, to a certain extent, some versions of the *simple distributed detection* (SDD) problem have efficient polynomial algorithms, there are versions of SDD that are computationally intractable; that is, no efficient solutions are known to exist.

There are different aspects of the distributed detection problem. We will not attempt to give an exhaustive coverage of the excellent papers in the literature. The section of selected readings give a list of references, while some of the problems require the reader to critique some papers.

We will cover Rao's treatment because it differs from the related works (e.g., *operational diagnostics*) by placing emphais on the design of efficient algorithms and data structures for the class of SDD problems that can be solved in polynomial time. Rao [62] also gives heuristic algorithms for the class of SDD problems that are hard; that is, cannot be solved in polynomial time.

If some SDD problems are hard, a logical question is, what makes such SDD Problems intractable? Rao, Iyengar and Kashyap [63] is a logical sequel to Rao [62]. They present the computational complexity of distributed detection problems with information constraints.

9.1 Simple Distributed Detection Problems

We will discuss simplified versions of the distributed detection problem. Our formulation closely follows Rao.

Consider a finite set of sensors S, $|S| = m$, for monitoring a work space. Assume that a finite set of objects O, $|O| = n$, can move in and out of the work space. Depending on the application domain, objects are also called targets or obstacles. For the moment, we further assume that at any given time, at most one obstacle can be present in the work space. This is called the single-object detection problem. Later, we will relax this restriction and allow more than one object to be present in the work space at any given time. This version is called the multiple-object detection problem.

Denote a representative sensor by a, (i.e., $a \in S$). A sensor can detect one or more objects. We will write $\text{Det}(a)$ to stand for the set of objects that can be detected by the sensor a. It is clear that $\text{Det}(a) \subseteq O$, where O is the set of all objects.

Denote a representative object by b, (i.e., $b \in O$). An object can be detected by one or more sensors. We will write $\text{Sen}(b)$ to stand for the set of sensors that can detect the object b. It is clear that $\text{Sen}(b) \subseteq S$, where S is the set of all sensors.

Now, we are ready to define the detection problem. Suppose an object (not necessarily from the set O) has been detected by one or more sensors. For each object, the system has to decide if it belongs to O. Furthermore, the system must return the set, say Ψ, of all objects that belong to O and that has been detected by one or more sensors. The set Ψ could possibly be empty.

Note that we have not yet defined the system. There are two obvious ways to define the system: one using the sensors and the other using the objects.

In the *forward* detection problem (FDP), we describe the system by $\{\text{Det}(a)\}_{a \in S}$; that is, for each sensor a, we give the set of objects it can detect (see Fig. 9.1).

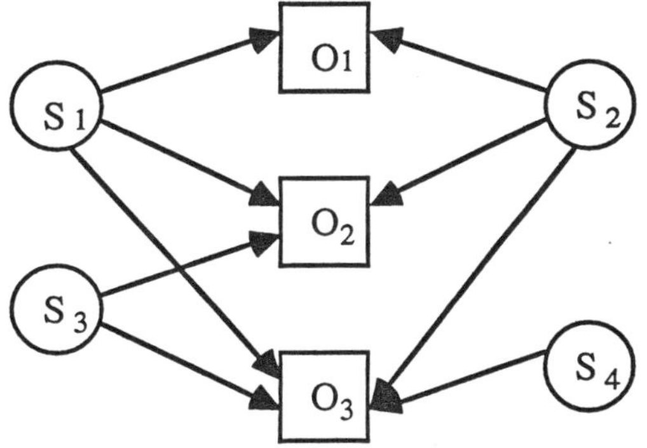

Sensor	Detects
S_1	$(\ O_1,\ O_2,\ O_3\ )$
S_2	$(\ O_1,\ O_2,\ O_3\ )$
S_3	$(\ O_2,\ O_3)$
S_4	$(\ O_3\ )$

Figure 9.1 Forward detection problem. $S = \{s_1, s_2, s_3, s_4\}$ is a set of four sensors. $O = \{o_1, o_2, o_3\}$ is a set of three objects. $\text{Det}(s_1) = \{o_1, o_2, o_3\}$ is the set of objects that can be detected by sensor s_1. $\text{Det}(s_2) = \{o_1, o_2, o_3\}$. $\text{Det}(s_3) = \{o_2, o_3\}$. $\text{Det}(s_4) = \{o_3\}$.

In the *backward* detection problem (BDP), we describe the system by $\{\text{Sen}(b)\}_{b \in O}$; that is, for each object b, we give the set of sensors that detect b (see Fig. 9.2).

Several questions come to mind? Which problem is easier (or harder) to solve, FDP or BDP? Could it be that FDP and BDP have the same complexity? Is it possible to convert FDP into BDP, and vice versa? These questions are the concerns of the following sections.

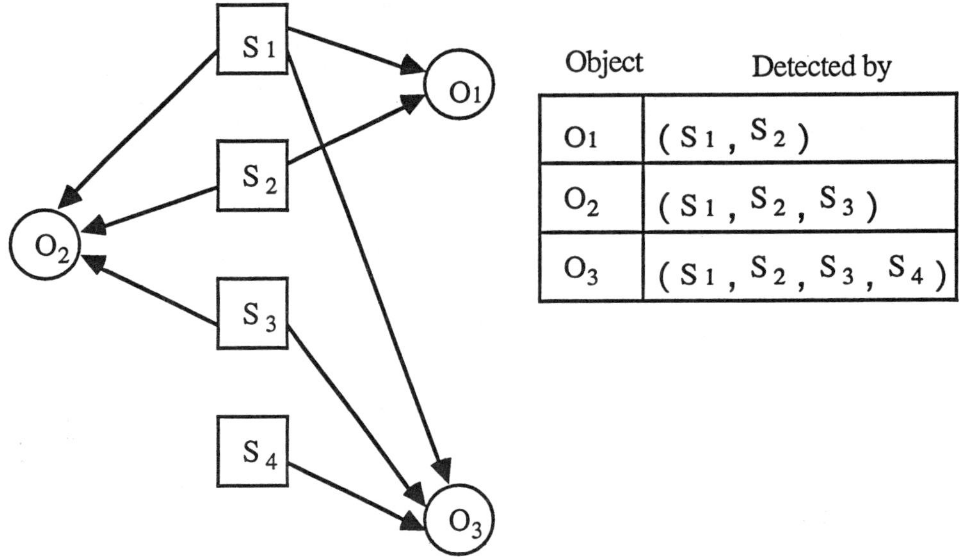

Object	Detected by
O_1	(S_1, S_2)
O_2	(S_1, S_2, S_3)
O_3	(S_1, S_2, S_3, S_4)

Figure 9.2 Backward detection problem. $S = \{s_1, s_2, s_3, s_4\}$ is a set of four sensors. $O = \{o_1, o_2, o_3\}$ is a set of three objects. $\text{Sen}(o_1) = \{s_1, s_2\}$ is the set of sensors that can detect object o_1. $\text{Sen}(o_2) = \{s_1, s_2, s_3\}$. $\text{Sen}(o_3) = \{s_1, s_2, s_3, s_4\}$.

9.1.1   Representation of Sets

Note that FDP is specified in terms of $\text{Det}(a)$ and BDP is specified in terms of $\text{Sen}(b)$. To solve the problems, we need efficient data structures and algorithms for manipulating sets and families of sets.

We are concerned mainly with finite sets. The set of sensors S has a cardinality of m. The set of objects O has a cardinality of n. Consider subsets of S and O. Let a representative subset be denoted by A. The context (of the problem) will make it clear if we are talking about sensors or objects.

In the preliminaries, we described that there are four ways to specify a set: (1) arbitrary set, (2) linearly ordered set, (3) a set with a range of integer values, and (4) natural numbers. In the preliminaries, we also discussed techniques for testing the equality of two sets. We are interested in the case where the two sets, A and B, are both subsets of the set S (or O).

9.1.2 Relating Forward and Backward Detection Problems

Suppose we have problems X and Y. If we know how to solve problem Y, and there is a way to transform problem X into problem Y, we say that X can be solved. This technique makes sense if problem Y can be solved efficiently and if problem X can be efficiently transformed into Y.

FDP can be transformed into BDP, and vice versa. *Conversion* maps FDP to BDP, and *inversion* maps BDP to FDP (see Figure. 9.3. Since FDP is specified using $\{\text{Det}(a)\}_{a \in S}$, a conversion algorithm must generate the corresponding $\{\text{Sen}(b)\}_{b \in O}$.

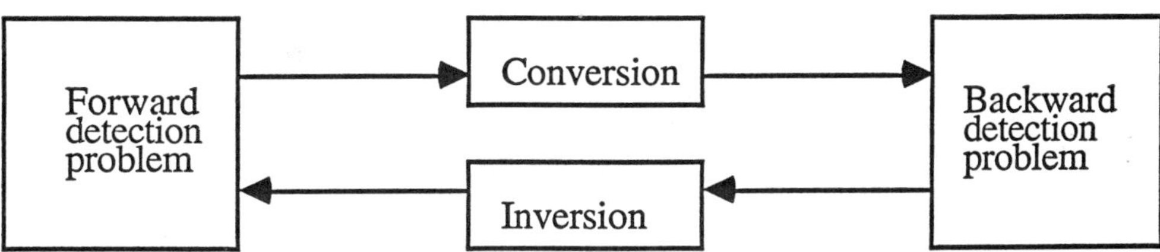

Figure 9.3 Conversion and inversion.

We will illustrate how the forward detection problem shown in Fig. 9.1 can be converted into the backward detection problem shown in Fig. 9.2. For each element $c \in \bigcup_{a \in S} \text{Det}(a)$, check its membership in each of the sets of $\{\text{Det}(a)\}_{a \in S}$. Return, as $\text{Sen}(c)$, the list of all a's such that $c \in \text{Det}(a)$.

The complexity of the *conversion* algorithm depends on the nature of the sets S and O. We will discuss the case where they are arbitrary sets. Start with a member of the first list. Check to see if it belongs to other lists as well. If so, delete it from each list in which it is contained. In our example, o_1 belongs to the first and second lists. We generate $\text{Sen}(o_1) = \{s_1, s_2\}$ and delete o_1 from both lists.

Repeat the process using another member of the first list. Since o_2 occurs in the first, second, and third lists, we generate $\text{Sen}(o_2) = \{s_1, s_2, s_3\}$ and delete o_2 from the respective lists. When all members of first list are exhausted, go to the second list and so on, until all lists are empty. For our example, we get $\text{Sen}(o_3) = \{s_1, s_2, s_3, s_4\}$. When S and O are arbitrary sets, the complexity of *conversion* is $O(n^2 m)$ because each $b \in O$ is checked for its membership in at most m sets, and there are at most n elements in each set.

In a similar manner, we can develop an *inversion* algorithm and analyze its complexity.

9.1.3   Detection Networks

A key idea is to translate the *detection process* into *reachability* of nodes in a detection network, which is a graph with the set of nodes $S \cup O$ such that there are no edges between nodes of S. The detection network $\xi(\{\mathrm{Sen}(b)\}_{b \in O})$ consists of the set of vertices $V - \xi$ and the set of edges E_ξ. A vertex (or edge) of the network is called a ξ-vertex (or ξ-edge).

A detection network has a *fundamental* property: There is a path from each node of $b \in O$ to every node $a \in \mathrm{Sen}(b)$ and there is no path from b to any node $S - \mathrm{Sen}(b)$. The fundamental property is also known as the *detection* property.

Given a detection problem, we can construct a detection network as follows.

Step 1: Rudimentary Graph

Construct a graph G as follows:

1. Make the first-level nodes of G correspond to the elements of O.

2. Make the second-level nodes of G correspond to the elements of S.

3. Connect each node $b \in O$ to each $a \in \mathrm{Sen}(b)$.

We will call G a rudimentary graph. G has at most nm edges. With respect to the detection property, any graph having more than nm edges is said to be superfluous and can be ignored.

Step 2: Detection Network

From G, there is a straightforward way to construct a detection network, $\xi(\{\mathrm{Sen}(b)\}_{b \in O})$. However, the number of ξ-edges might not be optimal.

Example 1: Given $S = \{1, 2, 3, 4\}$, $O = \{o_1, o_2, o_3\}$, $\mathrm{Sen}(o_1) = \{1, 2\}$, $\mathrm{Sen}(o_2) = \{1, 2, 3\}$, and $\mathrm{Sen}(o_3) = \{1, 2, 3, 4\}$. We construct a rudimentary graph G as follows. Create three level 1 nodes corresponding to the members of O. Create four level 2 nodes corresponding to the members of S. Connect the nodes directed by $\mathrm{Sen}(b)$. The nodes o_1, o_2 and o_3 to the have two, three and four edges respectively emanating from them; thus, there are nine ξ-edges in the straight forward detection network (see Fig. 9.4). The number of ξ-edges is not optimal.

A *nonsuperfluous* graph with the *detection* property satisfies the following:

1. Let $\xi(\{\mathrm{Sen}(b)\}_{b \in O}) = (V_\xi, E_\xi)$. $|V_\xi| \le (n + m)$ and $|E_\xi| \le nm$.

2. There is no path from a_1 to a_2 for $a_1, a_2 \in S$.

3. A path from b_1 to b_2, for $b_1, b_2 \in O$, implies $\mathrm{Sen}(b_2) \subseteq \mathrm{Sen}(b_1)$.

Ψ, the set of objects (in the work space) detected by the given set of sensors, can be computed from the detection network. *In-arborescence* of a ξ-node v, denoted by $IA(v)$, is the maximal set of ξ-nodes from which v is reachable.

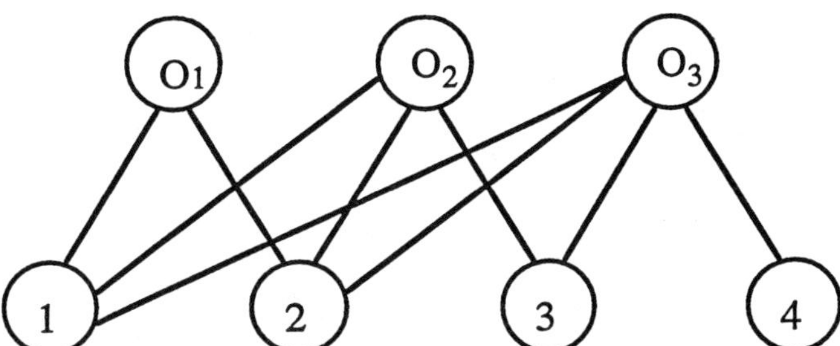

Figure 9.4 Straightforward detection network. Given $S = \{1, 2, 3, 4\}$ and $O = \{o_1, o_2, o_3\}$ such that $\mathrm{Sen}(o_1) = \{1, 2\}$, $\mathrm{Sen}(o_2) = \{1, 2, 3\}$, and $\mathrm{Sen}(o_3) = \{1, 2, 3, 4\}$. The network has nine ξ-edges and is not optimal.

Theorem 11 *For any graph structure that satisfies the detection property, we have*

$$\Psi = \bigcap_{v \in D} IA(v) - \bigcup_{v \in V_\xi - D} IA(v)$$

For any $v \in S$, we have $|IA(v)| \leq n + 1$ [62].

Proof: Any object $b \in O$ will be a potential candidate if and only if $\mathrm{Sen}(b) = D$, that is, $\Psi = \{b | b \in O \text{ and } \mathrm{Sen}(b) = D\}$. Consider $c \in \Psi$. By the detection property, there is a path from c to each member of D and there is no path to any member of $V_\xi - D$. Hence

$$\Psi \in \bigcap_{v \in D} IA(v) - \bigcup_{v \in V_\xi - D} IA(v).$$

A node c from

$$\bigcap_{v \in D} IA(v) - \bigcup_{v \in V_\xi - D} IA(v)$$

is definitely in Ψ. The bound on $IA(v)$ follows from the fact $IA(v)$ does not contain any sensor (possibly except for v) by property 2.

9.1.4 Computation of $\xi(\{\mathrm{Sen}(b)\}_{b \in O})$

In Fig. 9.4, we illustrated a nonoptimal detection network with nine ξ-edges. Notice that

$$\{1, 2\} \subseteq \{1, 2, 3\} \subseteq \{1, 2, 3, 4\}.$$

If we use multilevel trees (and property 3 of nonsuperfluous graphs with the detection property), we can make node o_1 a child of o_2, which in turn is made a child of o_1. This results in an optimal detection network with six ξ-edges as shown in Fig. 9.5.

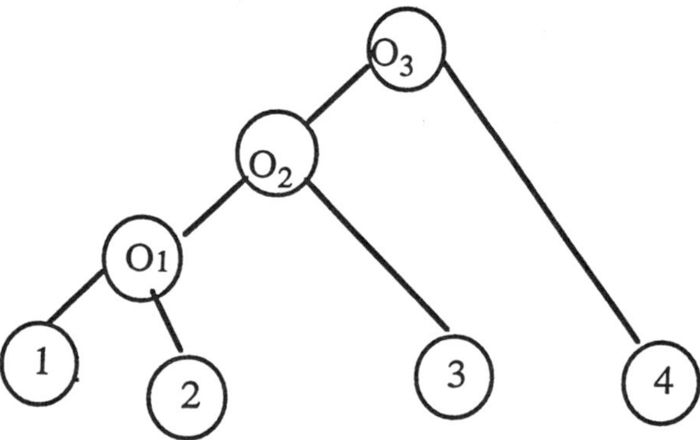

Figure 9.5 Optimal detection network. Given $S = \{1, 2, 3, 4\}$ and $O = \{o_1, o_2, o_3\}$ such that $\mathrm{Sen}(o_1) = \{1, 2\}$, $\mathrm{Sen}(o_2) = \{1, 2, 3\}$, and $\mathrm{Sen}(o_3) = \{1, 2, 3, 4\}$. Note that $\mathrm{Sen}(o_1) \subseteq \mathrm{Sen}(o_2) \subseteq \mathrm{Sen}(o_3)$. The network has six ξ-edges and is optimal.

Is it always possible to construct an optimal detection network, that is, $\xi(\{\mathrm{Sen}(b)\}_{b \in O})$, with a minimum number of ξ-edges? The answer is no, since the *minimal set covering* problem discussed in [21] and [38] can be reduced in polynomial time to the problem of an *optimal detection network problem*. We have to resort to an approximate algorithm to generate a possibly nonoptimal $\xi(\{\mathrm{Sen}(b)\}_{b \in O})$ [62].

Algorithm APPROX;

    Sort $\{\mathrm{Sen}(b)\}_{b \in O}$ according to the increasing values of $|\mathrm{Sen}(b)|$.
    Let $\{S_1, S_2, \ldots, S_n\}$ be the resultant sorted version.
    for i = 1 to n do
      begin
      $A = \{1, 2, \ldots, (i-1)\}$; $B = S_i$;
      while $k \in A$ exists, such that $S_k \in B$ **do**
      begin
      compute $S_j \in B$ such that $|S_j|$ is maximum;
      place an edge from S_i to S_j;
      $A = A - \{j\}$; $B = B - S_j$;

```
    end;
      place an edge from the node $S_i$ to every $a \in B$;
    end;
```

The APPROX algorithm can fail to give an optimal number of ξ-edges. Given $S = \{1, 2, 3, 4, 5, 6\}$, $O = \{o_1, o_2, \ldots, o_6\}$, $\text{Sen}(o_1) = \{1\}$, $\text{Sen}(o_2) = \{2\}$, $\text{Sen}(o_3) = \{1, 2, 3\}$, $\text{Sen}(o_4) = \{4, 5, 6\}$, $\text{Sen}(o_5) = \{3, 4, 5, 6\}$, $\text{Sen}(o_6) = \{1, 2, 3, 4, 5, 6\}$ (see Figure. 9.6.

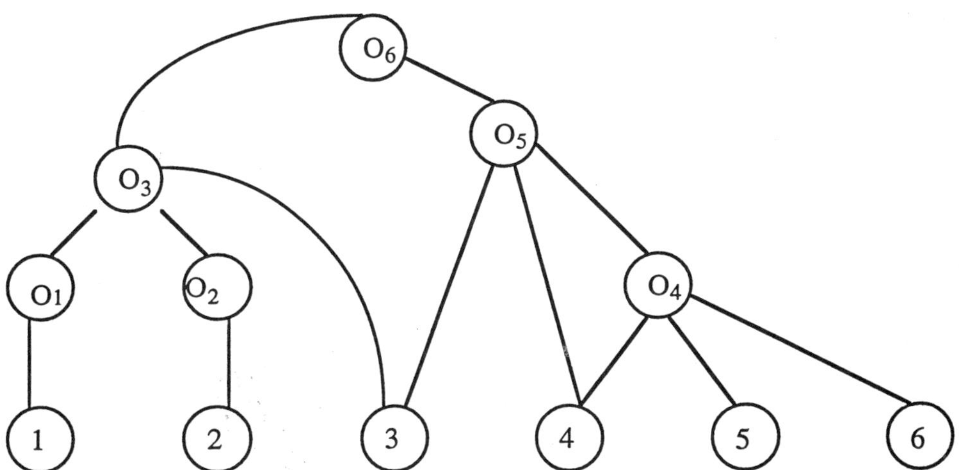

Figure 9.6 Given $S = \{1, 2, 3, 4, 5, 6\}$, $O = \{o_1, o_2, \ldots, o_6\}$, $\text{Sen}(o_1) = \{1\}$, $\text{Sen}(o_2) = \{2\}$, $\text{Sen}(o_3) = \{1, 2, 3\}$, $\text{Sen}(o_4) = \{4, 5, 6\}$, $\text{Sen}(o_5) = \{3, 4, 5, 6\}$, and $\text{Sen}(o_6) = \{1, 2, 3, 4, 5, 6\}$. The APPROX algorithm gives a non-optimal detection network with thirteen ξ-edges. The optimal detection network needs only twelve ξ-nodes.

The APPROX algorithm sorts the Sen's in increasing order of their cardinalities. It starts with $\text{Sen}(o_6)$ and recognizes that $\text{Sen}(o_5) \subseteq \text{Sen}(o_6)$. It makes o_5 the child of o_6. APPROX recognizes that $\text{Sen}(o_4) \subseteq \text{Sen}(o_5)$, and makes o_4 the child of

o_5. Even though $\text{Sen}(o_3) \subseteq \text{Sen}(o_6)$, APPROX cannot recognize the fact. Since $\text{Sen}(o_3) \not\subseteq \text{Sen}(o_4)$, APPROX creates a (sub)tree consisting of o_3, o_1 and o_2.

The corresponding optimal detection network shown in Fig. 9.7 has only 12 ξ-edges Notice that there is an edge from o_6 (root) to o_3 since $\text{Sen}(o_3) \subseteq \text{Sen}(o_6)$.

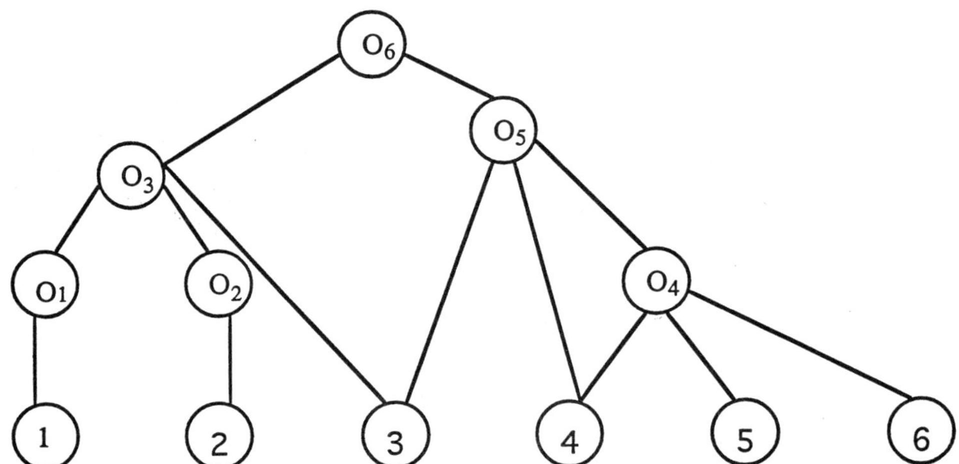

Figure 9.7 Given $S = \{1, 2, 3, 4, 5, 6\}$, $O = \{o_1, o_2, \ldots, o_6\}$, $\text{Sen}(o_1) = \{1\}$, $\text{Sen}(o_2) = \{2\}$, $\text{Sen}(o_3) = \{1, 2, 3\}$, $\text{Sen}(o_4) = \{4, 5, 6\}$, $\text{Sen}(o_5) = \{3, 4, 5, 6\}$, and $\text{Sen}(o_6) = \{1, 2, 3, 4, 5, 6\}$. The detection network has only twelve ξ-edges and is optimal.

Rao [62] gives another approximate algorithm for computing $\xi(\{\text{Sen}(b)\}_{b \in O})$ with minimum number of edges, based on an existing approximation algorithm [36] for the set covering problem.

9.1.5   Realization

The detection network can be realized using hardware or software. It can be implemented using combinational circuits, sequential computers (with or withour preprocessing), message passing systems, hypercubes, and parallel RAMs, to name a few. For a detailed discussion, see Rao [62].

9.1.6   Multiple-object Detection

Consider the case when multiple objects can be present in the work space. Assume that the objects do not occlude each other, i.e., every object present in the work space will be detected by the appropriate sensors. The idea is to identify the set of objects that is present in the work space using the sensor information. A decision version of the multiple-object recognition problem is posed as follows: Given that the sensors of D detect an object, does there exist a subset of objects $O' \in O$ of cardinality at most p such that $\cup_{b \in O'} \mathrm{Sen}(b) = D$. An answer of yes to this problem indicates that all the objects of O' are simultaneously present in the work space.

Theorem 12 *The decision version of the multiple object detection problem is NP-complete.*

Proof: This problem is in NP because any choice for O' can be checked to see if it is valid by computing

$$\left(\bigcup_{b \in O'} \mathrm{Sen}(b) \right)$$

We now illustrate a polynomial reduction from the *minimum set cover* problem to the present problem. Given an instance of the minimum set cover problem, generate an *instance* of the present problem as follows.

Construct a two level graph corresponding to this instance as follows.

$$O = \{i\}_{C_i \in C}$$

and

$$\mathrm{Sen}(b) = C_b, \forall b \in O$$

$$S = \bigcup_{C_i \in C} C_i$$

and $S = D$. With $k = p$, it is straightforward to see that a minimal cover exists if and only if suitable O' exists. Hence the theorem.

Note that the multiple-object detection becomes harder when occlusion of objects is possible because of the presence of the other objects. Rao's paper [62] essentially shows that even some simplified versions of the distributed detection problem can be intractable. The quest of what makes them intractable led to the study described in the next section.

9.2   Distributed Detection with Information Constraints

Consider the following system.

- A set of sensors $S = \{S_1, S_2, \ldots, S_m\}$

- A set of objects $O = \{O_1, O_2, \ldots, O_n\}$

- Information constraints are given by a relation $R \subseteq S \times O$ such that $(S_i, O_j) \in R$ if and only if S_i is capable of detecting O_j

- The sensor S_i produces an output of 1 or 0 when any object O_j, such that $(S_i, O_j) \in R$, has been sensed by S_i, and produces an output of 0 when O_j, such that (O_j, S_i) not in R, has been sensed.

- Each $(S_i, O_j) \in R$ is assigned a confidence factor which could be a numerical value, or a value that can be computed. Informally, the confidence factor indicates the likelihood of S_i producing an output of 1 when O_j has been sensed. In general, the confidence factor is expressed as a function $f : 2^O \times 2^S \to R^+$, where R^+ is the set of non-negative real numbers.

- $f(A, B)$, for $A \subseteq O, B \subseteq S$, denotes the confidence that A is the set of objects in the work space when the sensors B produce outputs of 1.

- It is assumed that $f(A, B)$ can be computed in $O(|A| + |B|)$ time.

- $\text{Sen}(O_j) = \{S_i | (S_i, O_j) \in R\}$

- $\text{Det}(S_i) = \{O_j | (S_i, O_j) \in R\}$.

Given that a subset of sensors $D \subseteq S$ has detected obstacles (i.e., produced an output of 1), the detection problem is to identify the subset of objects $H \subseteq O$ such that

$$f(H, D) = \max_{A \in 2^O} f(A, D)$$

that is, with maximum confidence value. Also, find the complexity of computing H.

Depending on the nature of the confidence factor f and the information constraints R, the computational complexity of the detection problem can be polynomial, NP complete, or NP hard.

The following simple versions of the detection problem are computationally intractable:

- Deterministic formulation, where confidence factors are either 0 or 1

- Uniform formulation where $(S_i, O_j) \in R$, for all $S_i \in S, O_j \in O$

- Decomposable systems under multiplication operation

The following versions are solvable in polynomial (in n) time:

- Single-object detection

- Probabilistically independent detection

- Decomposable systems under additive and nonfractional multiplicative measures

- Matroid systems

The simple detection systems of Rao [62], for which the detection is deterministic, are special cases of the present formulation. On the other hand, the probabilistic formulation of Demirbas [16] is a special case when $\text{Sen}(O_j) = S$, the confidence factor is the probability measure and the events (appearances of objects in the work space) are probabilistically independent. In terms of computational complexity, we show that this is a particularly easy case in that the detection can be carried out in $O(mn)$ time. Distributed detection problems based on probabilistic formulations have been extensively studied [10, 16, 66, 67, 77].

$f(\cdot)$ has lesser structure than a probability measure. As a special case, f can correspond to a probability measure.

The detection problem exhibits a myriad of complexity levels, ranging from a worst-case exponential (in n) to an $O(mn)$ time solvability. At one extreme, this problem has an exponential (in n) lower bound for the computational complexity; it is particularly interesting to note that it seems to be computationally harder than NP-complete problems. At the other extreme, it becomes solvable in $O(mn)$ time under additive measure. Some interesting subclasses, based on Bayesian-type methods (more specifically based on computing a posteriori probabilities), fall out as part of our analysis.

We are unware of systematic complexity studies of these classes of problems; we show that these problems are polynomial time solvable under probabilistically independent events, but become NP hard in a general case. The following simple versions of detection problems are computationally intractable (i.e., the optimization problems are NP hard and the decision versions are NP complete):

- Deterministic formulation, where the $f(\cdot)$ is interpreted as a probability measure on 2^O, and all detection probabilities are either 0 or 1

- Completely constrained system, where $\text{Sen}(O_i) = S$ for all $O_i \in O$

- f-decomposable and ff-decomposable systems under a product operation

The second case corresponds to the case of Bayesian detection when the events are not guaranteed to be independent.

The following versions are solvable in polynomial (in n) time:

- Single-object detection

- Probabilistically independent detection

- f-decomposable and ff-decomposable systems under additive measures and multiplicative measures, where $f(A) \geq 1$ for all $A \in 2^O$

- systems where the family of unions of $\text{Sen}(O_i)$ forms a matroid.

Note that the second case corresponds to the Bayesian detection where the events are considered independent.

A straight forward but inefficient solution to the detection problem can be obtained by explicitly computing $\{f(A, D)\}_{A \subseteq 2^O}$. The complexity $O(2^n(m + n))$ is prohibitively large.

The general detection problem has a lower bound of complexity of $\Omega(2^n)$, since it requires picking the largest of 2^n real numbers. However, the properties of f and R can be utilized to expedite the computation.

Consider the set of all unions of $\text{Sen}(O_j)$s given by

$$\xi = \{\text{Sen}(O_{i_1}) \cup \text{Sen}(O_{i_2}) \cup \ldots \cup \text{Sen}(O_{i_k}) | i_1, i_2, \cdots, i_n \in [1, n], k \in [1, n]\}$$

and note that $D \in \xi$.

For $A \in \xi$, let $\Gamma(A)$ denote the family of subsets of O such that each subset is related (under R) to the same subset A of S under the information constraints; that is,

$$\Gamma(A) = \{B | B \subseteq O, \bigcup_{a \in B} \text{Sen}(a) = A\} \ .$$

Now constrain the system such that $f(B_1, A) = f(B_2, A)$ for $B_1, B_2 \in \Gamma(A)$, i.e. for each set of sensors with output 1, all the sets of potential objects will have the same confidence, and the choice of any such set of objects constitutes a solution to the detection problem.

Consider the family $\Psi = \{\Gamma(A)\}_{A \in \xi}$. Thus, in this case we have

$$\max_{A \in 2^O} f(A, D) = \max_{A \in \Psi} f(A, D) \ .$$

If $|\Psi| \geq c2^n$ for some constant $c > 0$, the detection problem has a lower bound of $\Omega(2^n)$. However, if $|\Psi| \leq cp(n)$, for some constant $c \geq 0$ and some polynomial $p(n)$, the above argument cannot be applied to show an exponential lower bound.

We show that this problem is NP hard and contains several subproblems that are NP complete. Thus, this problem is still computationally intractable.

NP complete problems are the class of problems that can be solved in polynomial time on a nondeterministic Turing machine [21]. Informally, in these problems it can be verified that $A = H$ for any $H \subseteq O$ and for any $A \subseteq O$ in polynomial time, but there are exponential number of choices for A.

The nature of f and R is orthogonal in causing NP Completeness in certain subclasses of the detection problem. In particular either f or R could be damaging enough to make this problem intractable.

9.2.1   Multiple-object Detection

Consider the case where detection is deterministic in that confidence function $f(A, D) = 1$ if $\bigcup_{a \in A} \text{Sen}(a) = D$ and $f(A, D) = 0$ otherwise. Thus,

$$\max_{A \in 2^O} f(A, D)$$

corresponds to finding if there exists a subset $A \subseteq O$ that satisfies the information constraints imposed by R. This problem is called the multiple-object detection problem and is shown to be NP complete by Rao [62] by reducing the set cover problem to this problem. Thus the nature of information constraints R alone is sufficient to make this problem NP complete.

Some design problems are computationally similar to the multiple-object detection problem. For the case of single-object detection in a deterministic formulation, we may choose to ignore the outputs of some of the sensors. For example, we may like to compute a subset of sensors (if such a set exists) such that each object is detected by a unique set of sensors. This problem can be shown to the NP complete along the lines of the multiple-object detection problem.

9.2.2   Most Plausible Hypothesis

Consider a fully constrained system such that $\text{Sen}(O_i) = S$ for all $O_i \in O$. Let $f(\cdot)$ correspond to the probability measure on 2^O. Computation of H corresponds to computing a most plausible explanation under the set cover model. This problem has been shown to be NP hard by Reggia et al. [65].

Trivializing the nature of the information constraints R in the detection problem does not make it any easier computationally. In this case $|\Psi| = 1$, but the detection problem is computationally intractable. A polynomial bound on $|\Psi|$ does not make this problem solvable in polynomial time.

9.2.3   Decomposable Systems

A multiple-sensor system is f-decomposable if there exists an operator $\square$ such that for every $A, B \subseteq O$, $f(A \cup B, D) = \square(f(A, D), f(B, D))$, and $\square$ is computable in $O(|A| + |B| + |D|)$ time. Now notice that for any $A \subseteq O$, $f(A, D)$ can be computed in $O(|A|\log(|A|) + |D|\log(|A|))$ by using a straight-forward divide-and-conquer algorithm:

- Divide A into two sets A_1 and A_2 of approximately equal sizes.

- Recursively compute $f(A_1, D)$ and $f(A_2, D)$.

- Compute $\square(f(A_1, D), f(A_2, D))$.

The time complexity of this algorithm will be given by $T(|A|, |D|) = 2T(|A|/2, |D|) + O(|A| + |D|)$, which yields the desired complexity of $O(|A|\log(|A|) + |D|\log(|A|))$.

If $|\Psi| \geq c2^n$ for some $c > 0$, this problem will still have a $\Omega(2^n)$ lower bound. Thus, we are interested in the complexity of the detection problem for the case when $|\Psi|$ is polynomially bounded. If R is unconstrained, this problem subsumes the multiple-object recognition.

Consider the special case of a fully constrained system where $\square$ corresponds to multiplication of integers and $f(\cdot)$ is integer valued. We pose a decision version of the detection problem as follows: Given R and a positive integer b, is there

$H \subseteq O$ such that $f(H, D) = b$? This problem can be shown to be NP complete by establishing a polynomial time reduction from a well-known problem called the subset product problem, which is stated as follows [21]. Given a finite set A, size $s(a) \in Z^+$ for each $a \in A$, positive integer b, is there a subset $A' \subseteq A$ such that the product of the sizes of the elements in A' is exactly b? It is direct to see a reduction of this problem to the above version of the detection problem by identifying $f(\cdot)$ with the size function $s(\cdot)$.

Now consider that we can apply the divide-and-conquer algorithm on the set D also.

9.2.4 ff-Decomposable System

A multiple-sensor system is ff-decomposable if there exists an operator $\square$ such that for every $P, Q, R, T \subseteq O$,

$$f(P \cup Q, R \cup T) = \square(f(P, R), f(P, T), f(Q, R), f(Q, T))$$

and $\square$ is computable in $O(|P| + |Q| + |R| + |T|)$ time.

Now notice that for any $A \subseteq O$, $f(A, D)$ can be computed by using a straightforward divide-and-conquer algorithm. Now we have a recursive equation of the form

$$T(|A|, |D|) = 4T(|A|/2, |D|/2) + O(|A| + |D|) .$$

Let $|A| = 2^a$ and $|D| = 2^d$. For the case $a < d$, we obtain

$$T(|A|, |D|) = |A|^2[1 + T(1, |D|/|A|] + |A||D| .$$

Now we have

$$T(1, |D|/|A|) = O(|D|/|A|\log(|D|/|A|) = O(|D|/|A|\log(|D|) .$$

Thus, $T(|A|, |D|) = (|A|^2 + |A||D|[1 + \log(|D|)])$. Similar computation for the case $a > d$ together with this equation yields

$$T(|A|, |D|) = |A|^2 + |D|^2 + |A||D|\log(|A||D|)) .$$

Using the above discussion, we can show that the complexity of these systems is very similar to the f-decomposable systems.

9.2.5   Polynomial-time Solvable Classes

Single-object Detection

Assume no more than one object is present in the region monitored by the sensors. In this case, we compute $f(\{O_i\}, D)$ for each $O_i \in O$ and pick the object with highest $f(\cdot)$ value. This method has been used by Demirbas [15] in the special case where $f(\cdot)$ corresponds to the a posteriori probability computed from known a priori probabilities. The time complexity of this method is $O(nm)$. The same algorithm can be employed when it is given that no more than a constant number of objects could be present in the work space.

Completely Constrained Case

$f(\cdot)$ corresponds to the probability measures and the event of any obstacle being present in the work space is independent of any other being present. Note that we allow multiple objects to be present in the work space. Now we have $f(A \cup B, D) = f(A, D)f(B, D)$, where $A, B \in O$, where $f(X, D)$ is the probability that the objects $X \subseteq O$ are present in the work space.

$f(A \cup B, D) \leq \min(f(A, D), f(B, D))$ since each $f(\cdot)$ value is a fraction. Thus, we have

$$\max_{A \in 2^O} f(A, D) = \max_{O_i \in O} f(\{O_i\}, D) \ .$$

The latter can be computed in a straightforward manner in $O(nm)$ time. In other words, this problem reduces to that of single-object detection.

f-Decomposable Systems

Let the operation $\square$ correspond to addition on real numbers. In this case,

$$\max_{A \in 2^O} f(A, D) = f(O, D) \ .$$

that is, $H = O$.

A similar result applies when $\square$ is multiplication and $f(A) \geq 1$ for every $A \in 2^O$; but the problem becomes computationally intractable if no constraints are placed on $f(A)$s. Also, the above discussion holds when $\square$ could be dynamically switched between the addition and multiplication operations.

Matroid

We now state a definition of a matroid [55]. A subset system (E, Λ) is a finite set E together with a collection Λ of subsets of E closed under inclusion, that is, if $A \in \Lambda$ and $A' \subseteq A$, then $A' \in \Lambda$. The elements of Λ are called independent.

The combinatorial optimization problem associated with (E, Λ) is the following: Given $w(e) \geq 0$ for each $e \in \Lambda$, find an independent subset that has the largest possible total weight.

Matroids have special structure such that a greedy algorithm will yield a solution for the combinatorial optimization problem. See [55] for a detailed treatment of matroid-based algorithms. Now consider a f-decomposable system such that the subset system (S, ξ) is a matroid and the $\square$ of the decomposable system is addition. The detection problem in this case can be solved as follows. We first compute $f(\{O_i\}, \text{Sen}(O_i))$ for each O_i. Then we use the following greedy algorithm to compute H:

Algorithm :
begin
   $I := \Phi; H := \Phi; C := O;$
   while $(I \neq D)$ **do**

```
begin
  let O_k ∈ C such that
    f({O_k}, Sen(O_k)) = max_{O_i ∈ C} f({O_i}, Sen(O_i));
  if (I ∪ Sen(O_k) ⊆ D) then
  begin
    I := I ∪ Sen(O_k);
    H := H ∪ {O_k};
  end
  else
    C := C − {O_k};
  end;
  output H;
end
```

The correctness of this method follows by straightforward methods [55]. The computation of $f(\{O_i\}, \text{Sen}(O_i))$ for each O_i has a time complexity of $O(s)$. The complexity of the above procedure is $O(ns)$ since there are n iterations and each iteration can be carried out in $O(s)$ time.

9.3   References and Further Reading

Tsitsiklis and Athans [78] discuss the complexity of decentralized decision making and detection problems. The early parts of the chapter are adapted from the paper by Rao [62]. See the references cited there.

Rao shows that the problem of computing a detection network of optimum size is computationally intractable (NP hard). However, with the availability of an *efficient* detection network, the detection problem can be solved in lower-order polynomial times on several computing systems. By preprocessing the network structure, the detection process can be further expedited [62].

We later discussed the computational complexity of distributed detection with information constraints. The material is adapted from [63].

PROBLEM SET 9

9-1. What is the classical theory of detection (or sensor signal processing)? Can it be used for distributed sensor networks?

9-2. Tenney and Sandell [77] is a classic paper on distributed detection. Summarize the findings for the case when there are two local detectors.

9-3. What are the three common methods for fusing decisions in a distributed sensor network?

9-4. The paper by Geraniotis and Chau [22] discusses robust data fusion for multisensor detection systems. Give a critique of the paper.

9-5. Durrant-Whyte [17] discusses sensor models and multisensor integration. Highlight his approach.

9-6. Show that the forward detection problem (FDP) specified by cases (a), (b), and (c) can be reduced to case (d). Show the same for the backward detection problem. *Hint*: The complexities are given in Table 9.1.

Table 9.1 Complexity of Transforming cases (a) to (c) into case (d)

Case	Input given by $\{\text{Det}(a)\}_{a \in S}$	Input given by $\{\text{Sen}(b)\}_{b \in O}$
(a)	$O(n^2 m)$	$O(nm^2)$
(b)	$O(nm \log n)$	$O(nm \log m)$
(c)	$O(nm + mM)$	$O(nm + nM)$

9-7. Given two subsets, A and B, of the set S of sensors. What is the time complexity to check if they are equal? Do the same for the case when A and B are subsets of the set O of objects. *Hint*: The complexities are summarized in Table 9.2.

Table 9.2 Complexity of Checking Equality of Sets.

Case	Nature of A $A, B \in S$	Nature of B $A, B \in O$
(a)	$O(m^2)$	$O(n^2)$
(b)	$O(m \log m)$	$O(n \log n)$
(c)	$O(m + M)$	$O(n + M)$
(d)	$O(m)$	$O(n)$

9-8. Verify the following complexities for transforming a FDP into a BDP (see Table 9.3).

9-9. Verify that the algorithm APPROX generates a nonsuperfluous graph that satisfies the detection property. Also, verify the complexity given in Table 9.4.

Table 9.3 Complexities of Conversion and Inversion

Case	Conversion	Inversion
(a)	$O(n^2 m)$	$O(nm^2)$
(b)	$O(nm \log n)$	$O(nm \log m)$
(c)	$O(nm + mM)$	$O(nm + nM)$
(d)	$O(nm)$	$O(nm)$

Table 9.4 Complexity of the Algorithm APPROX

Case	Complexity
(a)	$O(n^2 m + nm^2)$
(b)	$O(n^2 m + nm \log m)$
(c)	$O(n^2 m + nM)$
(d)	$O(n^2 m)$

Chapter 10

Suggested Solutions

10.1 Suggested Solutions for Problem Set 2

2-1. Differentiate the terms data and information.
Suggested solution:
Datum (pl. data) is a fact, a figure, or a collection of facts and figures. Information is precise, accurate, timely, and useful data.

2-2. Describe the following: fail hard, fail soft, fail safe.
Suggested solution:
A system is said to be fail hard if a failure can cause it to break down completely. A system is said to be fail soft if its performance degrades gracefully in the presence of faults. A system is said to be fail safe if it has built-in redundancies to safeguard against disruption of service due to failures.

2-3. What is an information processing system? What kinds of processing systems are there?
Suggested solution:
An information processing system is a system that accepts raw input data and transforms them into useful data or *information*. According to some reckoning, computer-based processing systems perform

- Data processing

- Information processing

- Knowledge processing

- Intelligence processing

in increasing order of complexity.

10.2 Suggested Solutions for Problem Set 3

3-1. Animals possess a myriad of senses. Give some examples. Have they provided motivation for the development of intelligent sensor-based systems?

Suggested solution:

Bats and porpoises have built-in ultrasonic range finders. They are forerunners of radars. A rattlesnake is known to use both infrared and visual sensors to target prey. It is cited as an instance of multisensor fusion in nature. Dogs are noted for their hearing and smell. They are trained for use in police, army, navigation guides, and so on. Dogs can often outperform the million dollar synthetic bomb sniffers and security systems used in major airports.

3-2. Is the brain an inspiration for computing and communication paradigms?
Suggested solution:

The brain might have inspired the following paradigms:

- Distributed signal processing

- Distributed problem solving

- Fault-tolerant computing

- Control systems

- Artificial intelligence

- Computational intelligence (which covers neural nets, evolutionary algorithms, and fuzzy logic)

3-3. What is a feedback control system? Describe two feedback control systems that occur in nature.
Suggested solution:

A feedback control system consists of input, ouput, controller, and a set of sensors. The output of a process is sensed and fed back to the subtractor. The difference between the actual ouput and the desired output should be small and ideally zero. A nonnegligible difference, or error, causes a signal to be sent as feedback to the controller. The controller's action should ultimately result in a zero (or negligible) difference.

The two feedback control systems that occur in nature are homeostasis and the immune system. Homeostasis refers to the fact that the human system senses changes in body temperature and takes corrective action (e.g. sweating) to maintain it at a normal level.

The immune system can generally sense lapses in the defense system of a human body and produce antibodies to counterattack invaders. It has built-in sensors and biofeedback.

3-4. Attempts to understand and/or mimic human behavior have resulted in interesting machine systems and associated research. Can you name some?
Suggested solution:

Table 10.1 gives a sampling of such efforts.

Table 10.1 Sensors, Senses and Typical Machine Systems

Sensor	Sense	Typical Machine System
Eye	Sight	Computer Vision Image Processing
Ear	Hearing	Speech Recognition Music Synthesis
Nose	Smell	Bomb-sniffer
Skin	Touch	Prostheses
Tongue	Taste	Not known at the time of this writing

3-5. Give a macro/micro view of the human eye in the context of distributed sensor processing.

Suggested solution:

At the macro level, the two eyes are the sensors for human vision. Together, they provide a perception of depth (using parallax). They can discriminate colors. They can choose the level of details in a scene. Humans tend to look globally at a scene and then zoom in to look at the details. This focus/defocus is a basis of multiresolution techniques.

At the micro level, a human eye consists of about 120 million rod cells that are responsible for scotopic or night vision and approximately 6 million cone cells that are responsible for photopic or day vision. Two thousand cone cells are concentrated in the fovea and are responsible for seeing details in a scene. The eye is sensitive to color because different cones have different pigments: blue, green, and red.

The sensors are spatially distributed, but are not uniformly distributed. The disparate signals (e.g. color signals) have to be integrated in a meaningful manner by the brain.

3-6. Give examples of NMR redundancy.

Suggested solution:

A space shuttle has both hardware and software redundancy built into it. It has five computers; all are activated for critical periods like launch and reentry. It has two logically equivalent implementations of most software.

10.3   Suggested Solutions for Problem Set 4

4-1. What is a process control system?

Suggested solution:

A process control system consists of:

- A set of sensors (or inputs)

- A set of actuators (or outputs)

- An algorithm that accepts input from the sensors and takes corrective action (if required) on the actuators

The details of a process control system depend on the process as well as the type of control used.

4-2. Give an example of a process control system. What is the relevance of sensor integration in a process control system?

Suggested solution:

A typical process control system is shown in Fig. 10.1. A physical process is monitored using sensors $S_1, S_2, \ldots, S_N$. The output of the sensors form the input to a process control program (PCP), which checks the state of the physical process and decides what action is needed. The process control program writes output to the actuators 1 to M.

A simple example of a process control system is an environmental support system for a room housing several computers and workstations. Assume that the the room temperature must be maintained at $70° \pm 3°$ F and the relative humidity at 50 ± 0.5. There will be sensors for sensing the temperature and humidity of the room. Assume further that there are several thermometers in the system. The sensor outputs (i.e., temperature and relative humidity readings) are continuously monitored and sent to the process control system, which will decide the corrective action (say, set actuators).

Since there are multiple sensors in a process control system, sensor integration is needed to combine the different sensor readings. Marzullo suggested the use of abstract sensor estimates and proposed algorithms for fault-tolerant sensor integration of redundant abstract sensor estimates.

4-3. Suppose we are two sensors S_1 and S_2. At time t, their readings are $S_1 = [l_1, u_1]$ and $S_2 = [l_2, u_2]$. How can we integrate the readings if the two sensors overlap?

Suggested Solution

We are given that the two sensor readings overlap. Without loss of generality, assume $l_1 < l_2 < u_1 < u_2$ (see Fig. 10.2). Here N = total number of sensors = 2. F = faulty number of sensors = 0 or 1. We are required to find I_s, the smallest interval that is guaranteed to contain the correct physical value.

Case (i): $F = 0$ means both sensors are correct. Then I_s is given by

$$S_1 \cap S_2 = [l_1, u_1] \cap [l_2, u_2] = [l_2, u_1]$$

The *width* of the output $= u_1 - l_2$.

Case (ii): $F = 1$ means at most one sensor can be faulty. Then the correct value could be in one of the intervals $[l_1, u_1]$ or $[l_2, u_2]$. Since we do not know which one, we set $I_s = [l_1, u_2]$. The *width* of the ouput $= u_2 - l_1$.

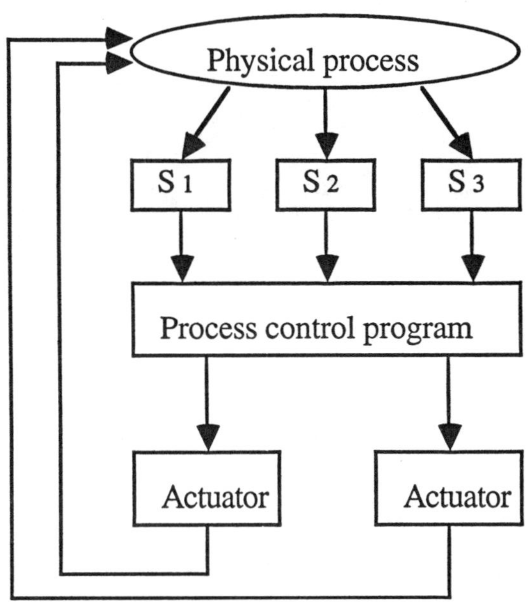

Figure 10.1 Process control system. The output of a physical process is monitored using n sensors. The process conrol program compares the integrated output against the desired output and activates the actuators (if necessary) to control the physical process.

4-4. For Problem 4.3, assume that $l_1 = 1.3$, $l_2 = 1.4$, $u_1 = 1.5$ and $u_2 = 1.6$. Find the width of the integrated output.

Suggested solution:

If there are no faulty sensors, the width of the integrated output = 1.5 - 1.4 = 0.1 If one sensor can be faulty, the width of the integrated output = 1.6 - 1.3 = 0.3.

Note that the presence of a faulty sensor can cause the width of the integrated output to increase.

4-5. Repeat Problem 4-3 for the case where the two sensor readings barely overlap. If there are slight perturbations in the sensor readings, would the integrated output remain the same?

Suggested solution:

We are given that the two sensor readings barely overlap. Without loss of

Figure 10.2 Integrating two sensors S_1 and S_2 where $l_1 < l_2 < u_1 < u_2$.

generality, assume $l_1 < l_2 < u_1 < u_2$. The situation is similar to Fig. 10.2 except that the values of l_2 and u_1 are nearly equal. Then $I_s = [u_1 - l_2]$ and the width is close to zero.

Slight perturbations might cause the values of l_2 and u_1 to change in such a way that the two sensor readings barely do not overlap. Without loss of generality, assume $l_1 < u_1 < l_2 < u_2$. The situation is similar to Figure 10.3 except that the values of u_1 and l_2 are nearly equal. Then, $I_s = [u_2 - l_1]$. The width is no longer close to zero.

This problem is intended to show the effect of imprecise measurements on the width of the integrated output. The width of the overlap of intervals should be comparable to the tolerance of the measurements; otherwise, the integrated output is not robust.

4-6. Draw interval graphs for the case where there are two, three and four sensors.
Suggested solution:

If there are only two sensors, the corresponding interval graph will have only two nodes. If the sensor readings overlap, we will draw an edge connecting the two nodes. If the sensor readings do not overlap, the nodes will remain isolated. Interval graphs can be drawn in a similar manner for three and four sensors.

4-7. Given three sensors S_1, S_2 and S_3. At time t, their readings are $S_1 = [l_1, u_1]$, $S_2 = [l_2, u_2]$, and $S_3 = [l_3, u_3]$. How can we integrate the sensor readings if all three sensor readings overlap?
Suggested solution:

We are given that all three sensors overlap. This implies that all are correct. Without loss of generality, assume $l_1 < l_2 < l_3 < u_1 < u_2 < u_3$. Then I_s is the intersection of the three intervals and is given by $[l_3, u_1]$ (see Fig. 10.4).

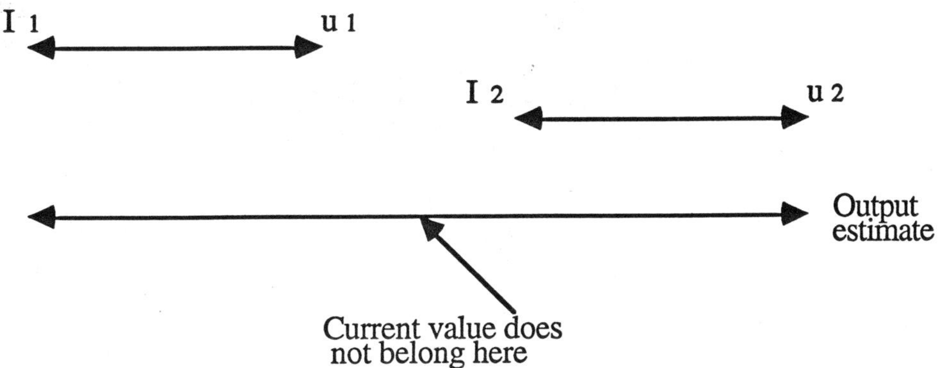

Figure 10.3 Interval graphs for two sensors S_1 and S_2. $l_1 < u_1 < l_2 < u_2$, the sensors do not overlap.

4-8. Repeat Problem 4-7 for the case where two sets of two sensor readings overlap.
Suggested solution:
 We are given that two pairs of sensors intersect. Without loss of generality, assume $l_1 < l_2 < u_1 < l_3 < u_2 < u_3$. Then there are two pairs of intersections. The corresponding intervals are $[l_2, u_1]$ and $[l_3, u_2]$ (see Fig. 10.5). The value of I_s depends on the value of F.

4-9. Repeat Problem 4-7 for the case where only one set of two sensor readings overlap.
Suggested solution:
 We are given that only one pair of sensors intersects. Without loss of generality, assume $l_1 < l_2 < u_1 < u_2 < l_3 < u_3$. Then there is only one intersection. The interval is $[l_2, u_1]$ (see Fig. 10.6).

4-10. Repeat Problem 4-7 for the case where all three sensor readings do not overlap.
Suggested solution:
 We are given that no pairs of sensors intersects. Without loss of generality, assume $l_1 < u_1 < l_2 < u_2 < l_3 < u_3$ (see Fig. 10.7). The value of I_s depends on the value of F.

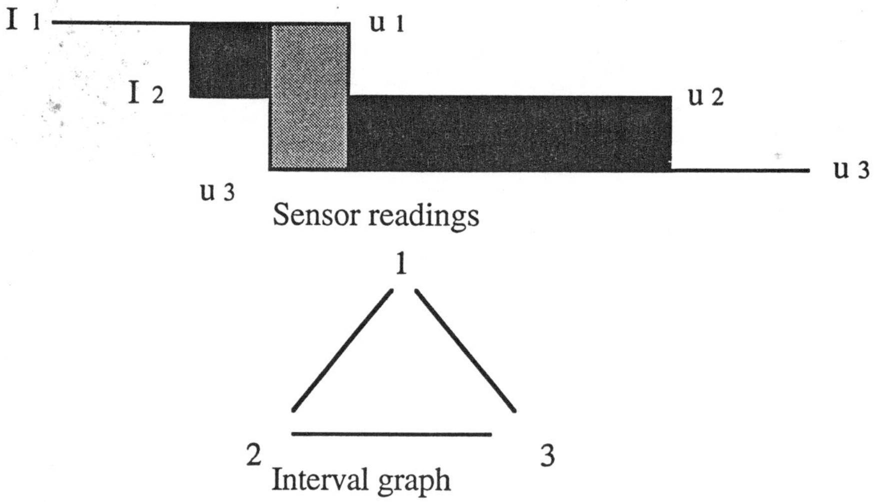

Figure 10.4 Integrating three sensors S_1, S_2, and S_3 where $l_1 < l_2 < l_3 < u_1 < u_2 < u_3$.

4-11. How can the integration of three sensors be applied to an intelligent node in a complete binary tree organization?

Suggested solution:

In a binary tree organization, there is a root node, several intermediate nodes, and more leaf nodes. The root node and the intermediate nodes are intelligent, that is, they have a processor for local processing.

The integration at the intelligent node will involve three sensor estimates:

1. Sensor reading at the parent node

2. Sensor estimate from the left child node

3. Sensor estimate from the right child node

4-12. Show that algorithm M1 is correct.

Suggested solution:

Algorithm M1 is a version of the fault-tolerant sensor averaging algorithm first proposed by Marzullo. Its proof of correctness can be found in Marzullo's paper [53]. An alternative proof is given in[60].

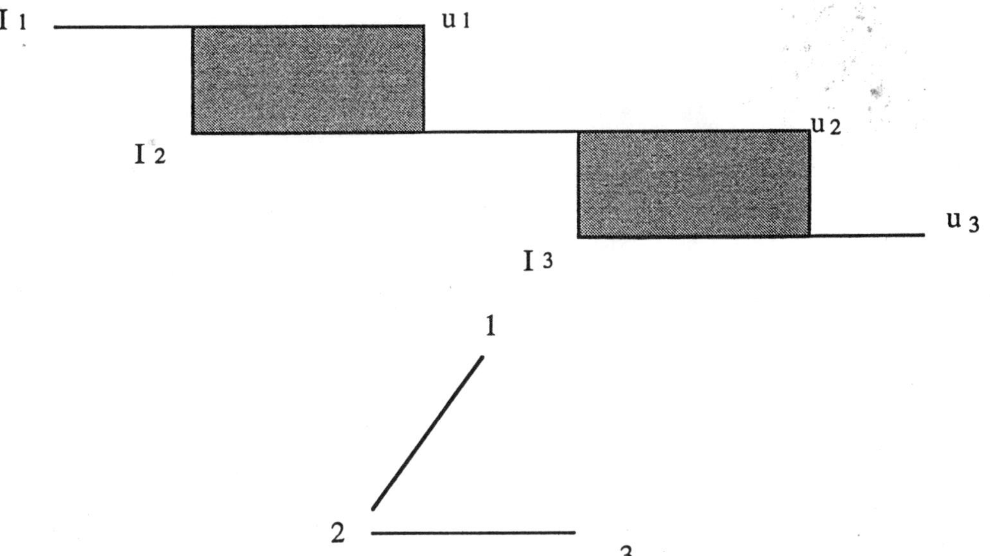

Figure 10.5 Integrating three sensors S_1, S_2, and S_3 where $l_1 < l_2 < u_1 < l_3 < u_2 < u_3$.

4-13. What is the union of intervals?
Suggested solution:
   The union of the intervals $A, B, \ldots, P$ is the interval $[\min(l_A, l_B, \ldots, l_P), \max(u_A, u_B, \ldots, u_P)]$.

4-14. What is a continguous interval?
Suggested solution:
   A contiguous interval is an interval made up of several intervals $A_1, A_2, \ldots, A_q$, where

$$u_{A_i} = l_{A_{i+1}}(1 \leq i \leq q - 1)$$

. $label_{A_i} \leq label_{A_{i+1}}$, and $l_{A_1}(u_{A_q})$ is the lower (upper) bound of C.

4-15. Implement algorithm J1 in a Pascal-like language.
Suggested solution:
   A version of Jayasimha's algorithm (presented in [9]) follows:

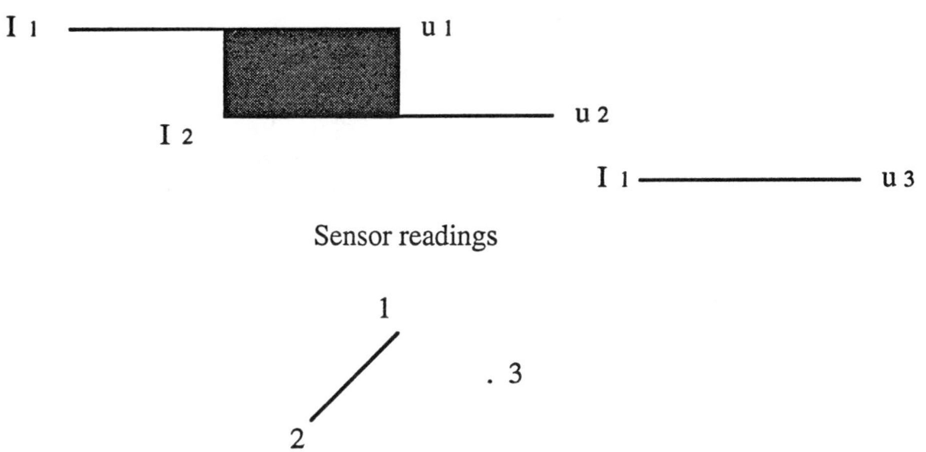

Figure 10.6 Integrating three sensors S_1, S_2, and S_3 where $l_1 < l_2 < u_1 < u_2 < l_3 < u_3$.

```
Jay_1D(DATA,i)
```

Input – DATA[N] a set of data in the form of a triplet:

  ID – Sensor identification number ID, which is unique for
each sensor and in the range 1 <= ID <= N.
  LOW[1..D] – Vector of lower bounds for each dimension.
  UP[1..D] – Vector of upper bounds for each dimension.

and i the dimension to be treated.

Output – LIST a list containing all the regions where >= (N-F)
    sensors intersect, this list has associated with it the set of
    members, and also the set of nonmembers.

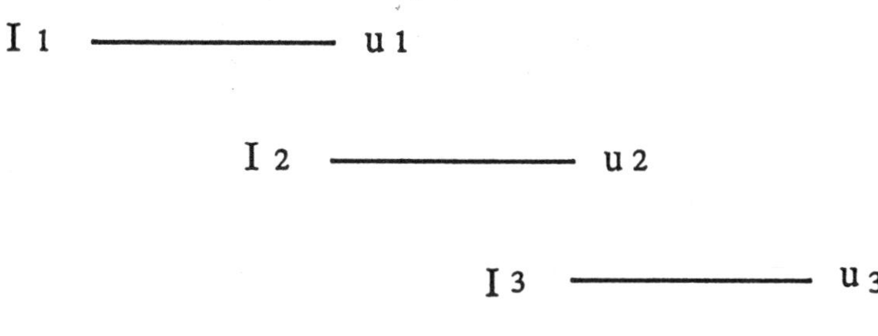

Figure 10.7 Integrating three sensors S_1, S_2 and S_3 where $l_1 < u_1 < l_2 < u_2 < l_3 < u_3$.

```
VAR low[N] /* Storage for lower bounds in sorted order */
 up[N-F-1] /* Storage for the N-F-1 largest upper bounds
 in sorted order */
 Clique /* Storage for clique to be put into LIST */

Begin
   Construct the array low[N] by sorting the lower bounds in
   increasing order for dimension i from DATA. Similarly construct
   array up[N-F-1] by sorting the upper bounds of the first
```

(N-F)-1 sensor readings from low.

LIST := NULL

for j := f downto 0 do
begin
Initialize Clique so that ID, range, members, and
nonmembers are all NULL.

Insert the upper bound from DATA associated with
low[N-j] into its appropriate position in array up.
Remove old values from the array if appropriate

If low[N-j] < up[1] then
begin
  The interval from low[N-j] to up[1] represents a
  clique with at least N-F members. Find the existing
  Cliques that intersect with this Clique and put
  them in a list C[1..P]

  Case 1: C := NULL
 Construct Clique with:
 Members := {sensors associated with values in the range
Low[N-j] to up[1]}
 Nonmembers := {sensors not included in Members}
 ID := Sum of characteristic numbers of the members of
  the clique.
 Range := low[N-j] to up[1]
 Add Clique to LIST

  Case 2: C <> NULL
 C[2..P-1] are all contained entirely in the region.
 for k := 2 to P-1 do
Add sensor associated with low[N-j] to Members set of C[k],
 remove it from Nonmembers set of C[k], and update the ID
 end /* For */

 if low end of range for C[1] = low[N-j] then
  begin
C[1] entirely contained in new range, so add sensor
associated with low[N-j] to Members set of C[1],
 remove it from Nonmembers set of C[k], and update
 the ID
  end

```
  else
begin
C[1] is not entirely in new range we split it
into two new regions. Modify upper value in
range of C[1] to low[N-j].
Construct a new Clique with:
Members := Members from C[1] and sensor associated
with low[N-j]
Nonmembers := Nonmembers from C[1] except for the
sensor associated with low[N-j]
ID := ID from C[1] + characteristic number of sensor
associated with low[N-j]
Range := low[N-j] to up[1]
Add Clique to LIST
  end
  if up[1] < upper limit of Range of C[P] then
 begin
 Split C[P] into two intervals. Modify lower range
 of C[P] to up[1]. Construct a new Clique with:
 Members := Members from C[P] and sensor associated
 with low[N-j]
 Nonmembers := Nonmembers from C[P] except for the
sensor associated with low[N-j]
 ID := ID from C[P] + characteristic number of sensor
 associated with low[N-j]
 Range := low[N-j] to lower limit of range of C[P]
 Add Clique to LIST
 end
 else
 begin
 C[P] is contained in new interval.
 Add sensor associated with low[N-j] to Members of
 C[P], Remove it from Nonmembers,
 and Update the ID number.
 end
  end
endfor
End.
```

10.4   Suggested Solutions for Problem Set 5

5-1. What is the overlap function $O(x)$ of the abstract sensors?
Suggested solution:
    The function $O(x)$ is the sum of the characteristic functions of the abstract

interval estimates. The value of $O(x)$ at any point x is the number of intervals overlapping at the point x.

5-2. Show that $O(x)$ has compact support.
Suggested solution:
There are a finite numbr of sensors, say N. Each sensor is represented by an abstract interval estimate of bounded length. Thus, $O(x)$ has compact support.

5-3. What is the time complexity for dynamically maintaining $O(x)$?
Suggested solution:
The maintenance of $O(x)$ requires $O(N \log N)$ time, where N is the number of sensors. This follows from the fact that sensor intervals need to be sorted first according to their beginning and end points, a process that has a computational complexity of $\Omega(N \log N)$.

5-4. What is an appropriate scale for performing multiresolution on $O(x)$?
Suggested solution:
The scale should not be too large or too small. Too large a scale will provide no useful information about the structure of O. Too small a scale would not isolate the features important to us, by bringing in unnecessary detail. Furthermore, since each sensor has width at least l, it is desirable to start off with a scale smaller than (or the same order as) l; that is, choose $j = \lceil \log(1/l) \rceil$. Thus, each sensor will figure as a feature at least as big as $1/2^j$.

5-5. What is a bitonic sequence?
Suggested solution:
A bitonic sequence is one increases first and then decreases. For details about bitonic sequences, see, for example [61]).

5-6. Suggest a way to construct random sensor intervals.
Suggested solution:
The correct-valued sensors can be constructed as

$$s_i = L - \omega \cdot \eta; \quad e_i = s_i + \omega, \tag{10.1}$$

where s_i is the start point of the interval, and e_i is the end-point of the interval. ω is the width of the sensor interval estimate and is constrained to lie between $\omega_{\min}$ and $\omega_{\max}$.

L in equation 10.1 is the correct value of the physical quantity estimated. η is a random number uniformly distributed over $(0.0, 1.0]$. For different sensors, ω is a random number uniformly distributed over $[\omega_{\min}, \omega_{\max}]$.

The intervals for tamely faulty sensors can be constructed as

$$s_i = L + t_{\text{off}} \cdot \eta; \quad e_i = s_i + \omega. \tag{10.2}$$

where t_{off}, a parameter of tame faults, denotes the critical distance from the correct value within which one of the end points of each tamely faulty sensor interval must lie.

Note that both the end-points of wildly faulty sensor intervals must lie at a distance of at least t_{off} from L. Intervals for wildly faulty sensors can be constructed as

$$s_i = L + t_{\text{off}} + \eta\,(L_{\text{max}} - t_{\text{off}} - \omega); \quad e_i = s_i + \omega, \tag{10.3}$$

where L_{max} is the maximum value any sensor can estimate. Faulty sensor intervals (both wild and tame) are located randomly on both sides of the correct value L.

5-7. What is the complexity of procedure *RESOLVE*?
Suggested solution:

Procedure *RESOLVE* involves only scanning and hence is linear in the number of sensors N. Since the average density of fluctuation in O is $2N/\text{Supp}(O(x))$, the level of resolution j required to capture almost all the measures of O is given by $j > \log(2N/\text{Supp}(O(x)))$.

If we assume that the parameter being measured by the sensors is known to lie between certain bounds, then $j > \log N + C$ where C is some constant. Thus procedure *RESOLVE* will be called, on an average, $O(\log N)$ times. Hence the average computational complexity of the algorithm is $O(N \log N)$.

10.5   Suggested Solutions for Problem Set 6

6-1. Fill in the implementation details in algorithm BI.
Suggested solution:

Algorithm BI follows.

```
Resolve-Data(DATA)


Input - DATA[N] A set of data from N independent abstract
sensors that measures a physical variable in Ddimensions. At
most F of the readings may be faulty. The data are in the form
of a triplet:

    ID - Sensor identification number ID, which is unique for
    each sensor and in the range 1 <= ID <= N.

    LOW[1..D] - Vector of lower bounds for each dimension.

    UP[1..D]  - Vector of upper bounds for each dimension.

Output - A Drectangle\index{Drectangle} containing the value of the physical
variable measured by the N sensors. This output is in the same
form as the sensor input:
```

FIN_LOW[1..D] -Vector of lower bounds for each dimension.

FIN_UP[1..D] -Vector of upper bounds for each dimension.

and FAULTY a list containing the set of all possible
combinations of faulty sensor readings from the input.

```
Var D          /* Number of dimensions the data uses */
    N          /* Number of sensors in the sensor array */
    Clique[D]  /* An array of lists containing all the cliques in
                  a given dimension */
    $F$           /* The maximum number of sensors may return
                  faulty readings and have the sensor array still
                  function correctly (<= N / 2D.) */
    SIZE[D]    /* for each dimension a list of cliques in descending
                  order of the number of members in the clique */

Begin

    For i := 1 to D do
       Solve the one-dimensional problems and get a list of
       all cliques of size >= N-F in that dimension (Clique[i])
       using the Jay_1D procedure
    end /* For */

    For i := 1 to D
       Construct SIZE[i] by sorting the CLIQUEs in ARRAY[i].LIST by
       descending size of the sets MEMBER.
    End /* For */

    For i := 1 to D
      For j := 1 to D
        If j <> i then
        begin

            Remove any CLIQUE from ARRAY[i].list that does not
            match at least one CLIQUE in SIZE[j]. A match occurs
            when one of the following is true:

          -There exists a clique of size N in SIZE[j]
           Lemma 1.

          -The number of members in SIZE[j] + the number
```

```
        of members in the CLIQUE under consideration
        is greater than or equal to 2N-F. Lemma 2.

      -There exists a CLIQUE in SIZE[j] where the
        size of the intersection between the elements
        not in the CLIQUE under consideration and a
        CLIQUE in SIZE[j] > 2N - (F + Sum of the
        number of elements in the two cliques)
        Lemma 3
        End
    End /* For */
End /* For */


For i := 1 to D do
    Search the values of RANGE for all CLIQUEs in
    Clique[i].LIST store:
        -The smallest lower bound in FIN_LOW[i]
        -The largest upper bound in FIN_UP[i]
End /* For */



Set FAULTY to an empty set
For i := 1 to D do
    Insert the contents of the set Nonmembers for each CLIQUE
    in ARRAY[i].list into FAULTY.
end /* For */


return(FIN_LOW, FIN_UP, FAULTY)


End.
```

6-2. Show that Algorithm BI is correct.

Suggested solution:

See the original paper by Brooks and Iyengar [9]. Note that algorithm BI requires three phases. The first phase uses a version of algorithm J1. Its proof of correctness can be found in [34].

An outline of the proof for Phase II follows. Once we have the set of regions or cliques where $N - F$ or more readings intersect for each dimension, we need to verify that the intersection exists in all dimensions. We do this by comparing our clique versus other cliques of $N - F$, or more, sensors in all other dimensions. If no possible match exists, then the intersection does not exist in at least one dimension and is disqualified.

6-3. Point out one potential problem with algorithm BI.

Suggested solution:

After Phase II of algorithm BI, the remaining regions consist of regions where $N - F$ sensors agree on the value of the physical variable in all dimensions.

This does not mean that there is always an intersection of $N - F$ sensor readings in all D dimensions. In cases of more than two dimensions, where there is significant clustering of sensor readings, it is possible for cliques that are not valid in all D dimensions not to be disqualified .

The principal drawback to the projection method is shown in a diagram presented in [13] similar to Fig. 6.2. The projection method treats the problem as D independent one-dimensional cases. Note that the shaded area in Fig. 6.2 is much larger in dimension 1 than the area actually covered by cliques of two members.

Figure. 6.3 shows the area returned by the algorithm presented here. It is of a higher time complexity than the projection algorithm, but returns a smaller region that is demonstrably correct.

An analogous problem does exist with the Algorithm, however. A clique with more than $(N - F)$ members may have intersections with two different $(N - F)$ member subsets in different dimensions. This clique would not be disqualified.

6-4. What is the efficiency of algorithm BI?

Suggested solution:

The One-Dim procedure requires $O(N \log N)$ time to execute. This is due to the sorting done at the start of the procedure. We therefore require $O(DN \log N)$ to perform the procedure for all dimensions. This matches the time complexity Chew Marzullo and [13] require for the projection algorithm.

Compare-Cliques has a worst-case complexity of $O(D^2 F^2)$ when bit vector operations, are elementary operations. If they are not elementary operations then the complexity is $O(D^2 F^3)$. This is arrived at using Jayasimha's [34] demonstration that at most $2F + 1$ regions exist for any one-dimensional case. We compare up to $2F + 1$ elements against at most $2F + 1$ elements for D dimensions versus $D - 1$ other dimensions. The complexity of the comparisons is explained in the discussion of lemmas 1 through 3.

Finding the boundaries of the Drectangle after we have eliminated the cliques that do not match cliques in other dimensions amounts to finding the minimum and maximum of D sets of $2F + 1$ elements. It has a complexity of $O(DF)$. Collecting the list of possible faults is trivial and can be done in $O(DF)$ time. The worst case complexity of the algorithm is thus $O(\max(DN \log N, D^2 F^2))$.

Note that the expected time complexity should be much smaller than this since any applications of Lemma 1 and 2 would greatly reduce the number of comparisons.

6-5. Discuss the optimality of algorithm BI.

Suggested solution:

Algorithm BI has optimal characteristics. The nature of the problem is such that other schemes that normally can be used to decrease algorithm complexity cause this problem to experience a combinatorial explosion.

The D^2 element of the complexity cannot be reduced since it is necessary to verify all dimensions against all other dimensions. A tournament scheme would actually increase the complexity of the algorithm. Comparing pairs of dimensions giving a list of cliques that are valid in both dimensions would usually decrease the number of cliques to be considered, but may increase the number of cliques, thus increasing the asymptotic complexity of the algorithm.

The F^2 element of the complexity cannot be reduced either. The number of cliques in each dimension may be up to $2F + 1$ [34]. With any method for ordering the cliques, it would be possible to construct an example for which the last clique in the ordering is the only valid clique for all dimensions. Our ordering, for which we start each validation with the clique with the largest number of members, is the one with the greatest chance of success for matching cliques and still may require up to F^2 comparisons.

Putting the cliques into tree structures based on the elements in the clique generally causes the complexity of the problem to become combinatorial as well.

6-6. Compare algorithm BI with other known methods.
Suggested solution:

Table 10.2 is a summary description of known algorithms. Algorithms J1 and BI algorithm also provide information as to which sensors might be faulty.

Table 10.2 Summary of Algorithms Discussed

Algorithm	Dim	Complexity	Region	Comments
Marzullo	1	$O(N \log N)$	Optimal	First algorithm
Jayasimha	1	$O(N \log N)$	Optimal	Lists faults
Chew and Marzullo If $n(s)$	2	$O(DN \log N)$	Optimal Drectangles	Prohibitive complexity for more than two dimensions
Chew and Marzullo Projection	D	$O(DN \log N)$	Drectangles larger than optimal	Ignores inter-dimensional effects
Brooks Iyengar	D	$O(\max(DN \log N, D^2 F^2))$	Drectangles frequently	Verifies inter dimensional effects

It is ambiguous as to whether such a clique should be disqualified or not. The case can be argued that data skewed in one dimension should be discarded in all other dimensions, this could also be a loss of potentially useful information [13]. It is our opinion that the compromise taken by this algorithm is reasonable, since any cliques not disqualified represent data returned by several sensors that concur in many (if not all) dimensions.

10.6   Suggested Solutions for Problem Set 7

7-1. What is a completely connected graph?
Suggested solution:
   A graph G is said to be completely connected if for any two vertices v_i and v_j there exists an edge (v_i, v_j). A completely connected graph with N nodes is denoted by K_N (see Fig. 7.2).

7-2. Give a formal description of a flat tree. Assume that each sensor cluster unit is a tree.
Suggested solution:
   A formal description of a sensor cluster unit graph is as follows: Let $s_1, \ldots, s_N$ ($N \geq 3$) be a set of processor/sensor nodes, or simply nodes. The graph for the sensor cluster unit composed of nodes $s_1, \ldots, s_N$ is a unidirectional acyclic graph C, which is a tree $T = \langle V, E \rangle$, where $V = \{s_1, \ldots, s_N\}$ is the set of all nodes in T and

$$E = \{(s_i, s_j) | s_i \text{ is the father node of } s_j, \text{ in } T\}$$

is the set of all edges (or links) in T. We require that the branch factor b of T be at least 2. The depth of the tree is $k \geq 1$, where k is called the level of the node. The root node of the tree T, s_r, is referred to as the commander of the sensor cluster unit, while the leaf nodes are referred to as low-level nodes.

   A formal description of a flat tree is as follows: Let $p_1, \ldots, p_M$ be a set of sensor cluster units. The graph for the flat tree DSN composed of sensor cluster units $p_1, \ldots, p_M$ is defined as a connected, undirected graph $G = \langle P, E \rangle$, where $P = \{p_1, \ldots, p_M\}$ is the set of all sensor cluster units nodes in G, and $E = \{(p_i, p_j)\}$ is the set of all edges in G. The nodes in P are connected by the edges in E, forming n sensor cluster unit subgraphs that contain M sensor cluster unit trees: $T_1, T_2, \ldots, T_M$, $(M \geq 2)$. For any two sensor cluster units i and j, the roots r_i, r_j are connected by an edge (r_i, r_j) in E. Thus all the root nodes and the edges among them form a completely connected subgraph of G.

7-3. Define the following: abstract sensor, interval, width of interval, intersection of intervals, interval inclusion, and distinct intervals.
Suggested solution:
   An abstract sensor is an abstraction of the reading of a concrete sensor (or physical device). It is generally represented by a *real interval R*. A *real interval*, or simply an *interval $R = (R_l, R_u)$* is represented by a pair of real numbers; R_l is called the *lower bound* and R_u is called the *upper bound* of the interval R. The *width* of the interval, $|R|$, equals $(R_u - R_l)$.
   The *set theoretic intersection* of two intervals, X and Y, is defined as $X \cap Y = \{c \mid c \in X \text{ and } c \in Y\}$ Two intervals are said to intersect (or overlap) if their set theoretic intersection is non empty. Hence, if the set theoretic intersection of X and Y is nonempty, then their interval intersection is the interval $(\max(X_l, Y_l), \min(X_u, Y_u))$.
   A special case of interval intersection is interval inclusion. X includes Y if

$X_l < Y_l$ and $X_u > Y_u$. The *span* of two intervals X and Y is $X \cap Y = (X_l, Y_u)$ Note that the span operation between two nonoverlapping intervals may result in an interval that includes points not lying in either of the intervals.

Intervals X and Y are said to be *non distinct* if either X includes Y or Y includes X; otherwise, X and Y are said to be distinct.

7-4. What are the properties of a deBruijn network?
Suggested solution:
A *de Bruijn network* has several good properties [68]:

- It possess good fault-tolerant capabilities.

- It can be configured as many useful computational networks in spite of faults.

- It has a small diameter.

- It allows simple routing.

10.7 Suggested Solutions for Problem Set 8

8-1. Name three regular tilings.
Suggested solution:
A plane can be tiled by three regular figures:

- Equilateral triangle

- Square

- Regular hexagon

8-2 How can we extend the concept of a correct sensor and a faulty sensor to the time-varying case?
Suggested solution:
An abstract sensor $\sum_{i,j}^{t} \in S_i$ is correct at $x \in E_i$ at time t if $(x, a) \in p_{i,j}^{t}(x)$, where a is the actual value of the parameter being measured at x at time t $(a \in P)$; else it is *faulty*.

10.8 Suggested Solutions for Problem Set 9

9-1. What is the classical theory of detection (or sensor signal processing)? Can it be used for distributed sensor networks?
Suggested solution:
The classical theory of sensor signal processing was developed for modeling the detection and tracking of targets using a single sensor (e.g., radar or sonar). It assumes that all the sensor signals are available at one place for processing. The theory uses statistical estimation and hypothesis testing methods.

The classical theory cannot be applied for use with a distributed sensor network. It is theoretically possible to transmit the sensor outputs from all sensors to the fusion center with negligible delay. However, in practice, a number of problems are associated with the disparate nature of the sensors, and total centralization of information rarely is feasible because of factors such as cost, reliability and communication bandwidth.

9-2. Tenney and Sandell [77] is a classic paper on distributed detection. Summarize the findings for the case when there are two local detectors.
Suggested solution:
Tenney and Sandell [77] consider the problem of detection using distributed sensors. They use decentralized optimal hypothesis testing.
For the case where there are two local detectors, they have shown the following:

- Each local detector implements a likelihood ratio (LR) test as its optimum decision rule.

- The computations of the thresholds for these tests are generally coupled.

- Locally optimal solutions to these computations can be achieved.

- Some solutions may run counter to intuition; for example, two identical detectors can have different thresholds even for symmetric cost functions.

- The theory applies to the case of decentralized detection with fusion through appropriate definition of a payoff function.

- Comparison of centralized and distributed decision rules shows a trade-off between communication and performance.

9-3. What are the three common methods for fusing decisions in a distributed sensor network?
Suggested solution:
Since a distributed sensor network consists of intelligent nodes, it is customary to let each node make binary decisions (e.g., whether the object has been detected) on the basis of its observations. The node then communicates the binary decisions to the fusion center, where the final decision must be made according to a rule.
Three commonly used fusion rules are block fusion, sequential fusion, and serial fusion.

Block fusion rule: The fusion center's decision test involves comparing to a threshold the likelihood ratio (LR) formed from the local decisions of (say K) sensors communicated to the fusion center in a fixed period of time, during which the local decisions are collected n times.

Sequential fusion rule: The fusion center executes a sequential probability ratio test (SPRT) based on the likelihood ratio (LR) formed from the local decisions of k sensors communicated to the fusion center until the current instant of time n (a random variable). If the likelihood ratio LR is above an upper threshold or below a lower threshold, a decision is reached by favoring hypothesis H1 or H0; if the LR is between the two thresholds, the local decisions of all k sensors are collected for the $(n + 1)$th time, a new LR is formed to include them, and the procedure is repeated.

Serial fusion rule: According to the serial fusion rule, the fusion center executes a sequential probability ratio test (SPRT) based on the likelihood ratio (LR) formed from the local decisions of the current number of sensors k (a random variable). If the LR is above an upper threshold or below a lower threshold, a decision is reached by favoring hypothesis H1 or H0, respectively; if the LR is between the two thresholds, the local decision of the $(K + 1)^{\text{th}}$ sensor is solicited, a new LR is formed to include it, and the procedure is repeated.

9-4. The paper by Geraniotis and Chau [22] discusses robust data fusion for multisensor detection systems. Give a critique of the paper.
Suggested solution:
Geraniotis and Chau [22] consider several problems of data fusion from sensors with observations experiencing statistical uncertainty. They have designed robust fusion rules, the (for the three known fusion rules - block fusion rule, sequential fusion rule, and serial fusion rule) and local decision schemes for the sensors, which guarantee acceptable performance levels despite the uncertainty. They have also evaluated the fusion rules when the number of sensors is large and when the observation interval is large in the fusion center.

9-5. Durrant-Whyte [17] discusses sensor models and multisensor integration. Highlight his approach.
Suggested solution:
Durrant-Whyte's model of a multisensor system represents the task environment as a collection of uncertain geometric objects. The sensors in the system are considered as a team of decision makers. Together, the sensors must determine a team consensus view of the environment. A multi-Bayesian approach with each sensor considered a Bayesian estimator is used to combine the associated probability distributions of each respective object into a joint posterior distribution function. A .likelihood function of this joint distribution is then maximized to provide the final fusion of the sensor information. Specifically, if the likelihood function $f_i(\cdot/P)$ is considered as a result of individual sensors (observers), then the final result is considered to be the joint posterior function $f(P/z_1, \ldots, z_n)$ after each sensor i in a n node distributed sensor network has made the observation z_i.

Appendix

Distributed Diameter-finding Algorithms

The following notations are used in describing *algorithm SDIA* and *algorithm MDIA*. $MSG[j]$ stands for the messaage sent by node j. $RCV(MSG[j]))$ stands for receipt of message sent by node j. i is used to represent an arbitrary node. s is used to represent a starting node.

Algorithm SDIA

Algorithm SDIA essentially consists of two kinds of algorithms: (1) when node i is not a start node, and (2) when node i is a start node.

Algorithm SDIA :

variables at node F_i:
```
N[i] = set of neighbors of node i.
d[1], d[2] = integers, initially 0.
lpd = diameter, initially 0.
state[i] = state of the node, initially idle.
SF[i] = sender of find message, initially nil.
RC = return message count, initially 0.
```

Algorithm at node i (different from start node s):

```
when (state[i] == idle)
  upon RCV(FIND[j])
    if (i == leaf node)
    then
      send(RETURN(0,0)) to j;
      exit
    else
      send(FIND) to all k, k in N[i]j;
      SF[i] = j;
```

      state[i] = *wait*;

when (state[i] == *wait*)
  upon RCV(RETURN(d, p[d]) j)
    d = d + 1;
    RC = RC + 1;
    compute d[1] and d[2] as the two largest d's received.
    lpd = max(lpd, p[d], d[1] + d[2]);
    if (RC == |N[i]| - 1)
    then
    send(RETURN(max(d1, d[2]), lpd) to SF[i];
    exit
end {algorithm of nonstart node}

Algorithm at node s, a start node.

  Send(FIND) to all k in N[s];
  state[s] = *wait*;

when (state[s] == *wait*)
  upon RCV(RETURN(d, p[d]) j)
    d = d + 1;
    RC = RC + 1;
    compute d[1] and d[2] as the two largest d's received.
    lpd = max(lpd, p[d], d[1] + d[2]);
    if (RC == |N[s]|)
    then
      lpd = diameter of the tree;
      exit
end {algorithm of start node}

Algorithm MDIA

The variables used in *MDIA* are the same as in *SDIA* except that *RC* is a set
variable.
Algorithm MDIA :

Algorithm at node i:

  when(state[i] == *idle*)
    upon RCV(FIND[j]) or upon START
    state[i] = *active*;
    if(i == leaf node)
    then

```
      send(RETURN(0,0)) to j;
   else
      SF[i] = j if i is not a start node;
      send(FIND) to all k, k in N[i]  SF[i];


upon RCV(RETURN(d, p[d]) j)
{ some leaf node started the algorithm }


d = d + 1;
RC = RC ∪ {j};
compute d[1] and d[2] as the two largest received d's.
lpd = max(p[d], lpd, d[1]+d[2]);

if( RC == |N[i]| )
then
   lpd is the tree diameter
   state[i] = terminal;
   send(SETDIA(lpd)) to all k, k in N[i];
   exit
else
if( |RC| - |N[i]| == 1)
then
   RETURN received from all but one neighbor
   state[i] = active;
   send(RETURN(max(d[1], d[2]), lpd);
else
   state[i] = active;
   send(FIND) to all k, k in i ; N[i]  j;

when(state[i] == active)
   upon RCV(FIND[j])
      ignore message

   upon RCV(RETURN(d, p[d]) j)
   { some leaf node started the algorithm }
   d = d + 1;
   RC = RC ∪ {j};
   compute d[1] and d[2] as the two largest received d's.
   lpd = max(p[d], lpd, d[1]+d[2]);

   if( RC == |N[i]| )
   then
      lpd is the tree diameter
      state[i] = terminal;
```

```
      exit
   else
   if( |RC| - |N[i]| == 1)
   then
      RETURN received from all but one neighbor
      state[i] = active;
   else
      state[i] = active;

   upon RCV(SETDIA(pd), j)
      lpd = pd;  diameter of the tree ;
      send(SETDIA(lpd)) to all k, k in N[i]  j.
      exit
end{algorithm}
```

Bibliography

[1] M. A. Abidi and R. C. Gonzalez, editors. *Data Fusion in Robotics and Machine Intelligence*. Academic Press, New York, 1992.

[2] A. V. Aho, J. E. Hopcroft, and J. D. Ullman. *The Design and Analysis of Computer Algorithms*. /newblock Addison-Wesley, Reading, MA, 1974.

[3] Y. Aloimonos and A. Rosenfeld. Computer Vision. *Science*, 253:1249–1254, September 1991.

[4] A. E. Baratz and J. M. Jaffe. Establishing virtual circuits in large computer networks. *Computer Networks*, 12:27–37, 1986.

[5] P. J. Besl. Active, optical range imaging sensors. *Machine Vision and Applications*, pages 127–152, 1988.

[6] U. K. Bhargava, R. L. Kashyap, and F. M. Goodman. Two nonparametric techniques for impulse response identification. *IEEE Trans. ASSP*, pages 984–995, July 1987.

[7] J. Bond and C. Pyerat. Diameter vulnerability in networks. In Y. Alavi et al., editor, *Graph Theory with Application to Algorithms and Computer Science*, pages 125–149. John Wiley and Sons, New York, 1985.

[8] J. M. Brady. Foreword. *Int. J. Robotics Research*, 7(5):2–4, 1988. Special Issue on Sensor Data Fusion.

[9] R. R. Brooks and S. S. Iyengar. Averaging algorithm for multi-dimensional redundant sensor arrays: Resolving sensor inconsistencies. submitted for publication, 1994.

[10] Z. Chair and P. K. Varshney. Optimal data fusion in multiple sensor detection systems. *IEEE Trans. Aerospace Electronic Syst.*, 22(1):98–101, 1986.

[11] C. Chatterjee, R. L. Kashyap, and G. Boray. Estimation of close sinusoids in colored noise and model discrimination. *IEEE Trans. on ASSP*, pages 328–337, March 1987.

[12] Y. Cheng and R. L. Kashyap. An axiomatic approach for combining evidence from a variety of sources. *J. Intelligent Robotic Systems*, March 1988.

[13] P. Chew and K. Marzullo. Masking failures of multidimensional sensors. In *Proc. 10th Symposium on Reliable Distributed Systems*, Oct. 1991. Pisa, Italy.

[14] I. Cidon and R. Rom. Fail-safe end-to-end protocols in computer networks with changing topology. *IEEE Trans. Communications*, COM-35(4):410–413, April 1987.

[15] K. Demirbas. Maximum a posteriori approach to object recognition with distributed sensors. *IEEE Trans. Aerospace Electronic Systems*, AES-24(3):309–313, 1988.

[16] K. Demirbas. Distributed sensor data fusion with binary decision trees. *IEEE Trans. Aerospace Electronic Systems*, AES-25(5):643–649, Sept. 1989.

[17] H. F. Durrant-Whyte. Sensor models and Multisensor Integration. *I. J. Robotics Research*, 7(6):97–113, 1988.

[18] A. H. Esfahanian and S. L. Hakimi. Fault-tolerant routing in de Bruijn communication networks. *IEEE Trans. Computers*, C-34(9), 1985.

[19] C. Esposito, D. Sullivan, U. frank, and R. Cibulskis. Field applications of robotic systems in hazardous waste site operaions. In M. Jamshidi and P. J. Eicker, editors, *Robotics and Remote Systems for Hazardous Environments*. Prentice Hall, Englewood Cliffs, New Jersey, 1993.

[20] M. J. Flynn. Some computer organizations and their effectiveness. *IEEE Trans. on Computers*, C-21:948–960, Sept. 1972.

[21] M. R. Garey and D. S. Johnson. *Computers and Intractability: A Guide to the Theory of NP Completeness*. W. H. Freeman, San Francisco, 1979.

[22] E. Geraniotis and Y. A. Chau. Robust data fusion for multisensor detection systems. *IEEE Trans. Information Theory*, 36(6):1265–1279, Nov. 1990.

[23] M. C. Golumbic. *Algorithmic Graph Theory and Perfect Graphs*. Academic Press, New York, 1980.

[24] B. Grofman and G. Owen, editors. *Information Pooling and Group Decision Making*. Jai Press Inc., Greenwich, Connecticut, 1986.

[25] R. Gusella and S. Zatti. The accuracy of the clock synchronization achieved by TEMPO in Berkeley UNIX 4.3 BSD. *IEEE Trans. Software Engineering*, pages 847–853, July 1989.

[26] T. C. Henderson and E. Weitz. Multisensor integration in a multiprocessor environment. In R. Raghavan and T. J. Cokons, editors, *Proc. ASME International Computers in Engineering Conference and Exhibition*, pages 311–316, New York, Aug. 1987.

[27] S. S. Iyengar, D. N. Jayasimha, and D. Nadig. A versatile architecture for distributed sensor integration problem. *IEEE Trans. on Computers*, Mar. 1994.

[28] S. S. Iyengar, R. L. Kashyap, R. N. Madan, and D. Thomas. A tree architecture for sensor fusion problems. In *Proc. of SPIE, Technical Symposium on Sensor Fusion*, April 1990. Orlando, Florida.

[29] S. S. Iyengar and Parameswaran. Autonomous intelligent systems. In preparation, 1995.

[30] S. S. Iyengar, M. B. Sharma, and R. L. Kashyap. Information routing and reliability issues in distributed sensor networks. *IEEE Trans. Signal Processing*, March. 1992.

[31] S. S. Iyengar and D. Thomas. A distributed sensor network structure with fault-tolerant facilities. In *Proc. of SPIE Symposium on Advances in Intelligent Systems*, Nov. 1989.

[32] J. M. Jaffe, A. E. Baratz, and A. Segall. Subtle design issues in the design of distributed dynamic routing algorithms. *Computer Networks*, 12(1):146–158, Aug. 1986.

[33] M. Jamshidi and P. J. Eicker. *Robotics and Remote Ssytems for Hazardous Environments*. Prentice Hall, Englewood Cliffs, New Jersey, 1993. Environmental and Intelligent Manufacturing Systems Series.

[34] D. N. Jayasimha. Fault tolerance in a multisensor environment. Technical report, Department of Computer Science, Ohio State University, 1993. Fall.

[35] D. N. Jayasimha, S. S. Iyengar, and R. L. Kashyap. Information integration and synchronization in distributed sensor networks. *IEEE Trans. Systems, Man, Cybernetics*, 21(5):1032–1043, 1991.

[36] D. S. Johnson. Approximation algorithms for combinatorial problems. *J. Computer System Sciences*, 9:256–278, 1974.

[37] T. T. Kadota. Integration of complementary detection-localization systems — an example. *IEEE Trans. Information Theory*, 40:808–819, 1994.

[38] R. M. Karp. Reducibility among combinatorial problems. In R. E. Miller and J. W. Thatcher, editors, *Complexity of Computer Communications*, pages 85–104. Plenum, New York, 1972.

[39] R. Kasturi and R. C. Jain. *Computer Vision: Advances and Applications.* IEEE Computer Society Press, Los Alamitos, CA, 1991.

[40] R. Kasturi and R. C. Jain. *Computer Vision: Priniciples.* IEEE Computer Society Press, Los Alamitos, CA, 1991.

[41] A. Kleinrock and F. Karmoun. Hierarchical routing for large networks – Performance evaluation and optimization. *Computer Networks*, 1(3):155–174, 1977.

[42] D. E. Knuth, J. H. Morris, and V. R. Pratt. Fast pattern matching in strings. *SIAM J. Computing*, 6:323–350, 1977.

[43] E. Korach, D. Rotem, and N. Santoro. Distributed algorithm for finding centers and medians in networks. *ACM Trans. on Programming Languages and Systems*, 6(3):380–401, 1984.

[44] K. B. Lakshmanan, K. Thulasiraman, and M. A. Comeau. An efficient distributed protocol for finding shortest paths in networks with negative weights. *IEEE Tran. Software Engineering*, 15(5):639–644, 1989.

[45] L. Lamport. Synchroned time servers. Technical Report DEC SRC TR, Digital Equipment Corporation, June 1987.

[46] R. F. Lax. *Modern Algebra and Discrete Structures.* Harper-Collins, 1991.

[47] V. R. Lesser, D. D. Corkill, J. Pavlis, L. Leikowitz, E. Hadlicka, and R. Brooks. A high level simulation test bed for cooperative distributed problem solving. In *Proc. DSN Workshop*, pages 247–270, 1982. MIT Lincoln Lab.

[48] T. Li and I. K. Sethi. Disributed decision fusion in the presence of link failures. To appear in IEEE Trans. Aerospace Electronics Systems.

[49] T. Li and I. K. Sethi. Optimal multiple level decision fusion with distributed sensors. *IEEE Trans. on Aerospace Electronic Systems*, 29(4), 1993.

[50] R. C. Luo and M. G. Kay. Multisensor integration and fusion in intelligent systems. *IEEE Trans. Systems, Man, Cybernetics*, 19(5):901–927, 1989.

[51] R. C. Luo and M. G. Kay. Data fusion and sensor integration: State-of-the-art 1990s. In M. A. Abidi and R. C. Gonzalez, editors, *Data Fusion in Robotics and Machine Learning*, pages 7–135. Academic Press, New York, 1992.

[52] R. C. Luo, M. Lin, and R. S. Scherp. Dynamic multisensor data fusion system for intelligent robots. *IEEE J. Robotics, Automation*, RA-4:386–396, 1988.

[53] K. Marzullo. Tolerating failures of continuous-valued sensors. *ACM Trans. Computer Systems*, 4(4):284–304, 1990.

[54] M. E. Merlin and A. Segall. A fail-safe distributed routing protocol. *IEEE Trans. Communications*, COM-27(9):1280–1287, 1979.

[55] C. H. Papadimitriou and K. Steiglitz. *Combinatorial Optimization Algorithms and Complexity*. Prentice-Hall, Englewood Cliffs, NJ, 1982.

[56] D. K. Pradhan. Interconnection topologies for fault-tolerant parallel and distributed architectures. In *Proc. 10th Int. Conf. Parallel Processing*, pages 238–242, Aug. 1981.

[57] D. K. Pradhan. Dynamically restructurable fault-tolerant processor network architecture. *IEEE Trans. Computers*, C-34(5), May 1985.

[58] D. K. Pradhan. Fault-tolerant VLSI architectures based on de Bruijn graphs (Galileo in the mid nineties). *DIMACS Series in Discrete Mathematics and Theoretical Computer Science*, 5, 1991.

[59] D. K. Pradhan and S. M. Reddy. A fault-tolerant communication architecture for distributed systems. *IEEE Trans. Computers*, C-31(9), 1982.

[60] L. Prasad, S. S. Iyengar, R. L. Kashyap, and R. N. Madan. Functional characterization of sensor integration in distributed sensor networks. *IEEE Trans. SMC*, 21(5), 1991.

[61] M. J. Quinn. *Designing Efficient Algorithms for Parallel Computers*. McGraw Hill, New York, 1987.

[62] N. S. V. Rao. Computational complexity issues in synthesis of simple distributed detection networks. *IEEE Transactions on Systems, Man, Cybernetics*, 21(5):1071–1081, 1991.

[63] N. S. V. Rao, S. S. Iyengar, and R. L. Kashyap. Computational complexity of distributed detection problems with information constraints. *Computers and Electrical Engineering*, 19(6):445–451, 1993.

[64] M. Raynal. *Networks and Distributed Computation: Concepts, Tools and Algorithms*. MIT Press, Cambridge, MA, 1988.

[65] J. A. Reggia, D. S. Nau, P. Y. Wang, and Y. Peng. A formal model of diagnostic inference – problem formulation and decomposition. *Inform. Sciences*, 37:227–256, 1985.

[66] A. R. Reibman and L.W. Nolte. Optimal detection and performance of distributed sensors systems. *IEEE Trans. Aerospace Electronics Syst.*, AES-23(1):24–30, 1987.

[67] F. A. Sadjadi. Hypothesis testing in a distributed environment. *IEEE Trans. Aerospace Electronic Systems*, AES-22:134–137, 1986.

[68] R. M. Samantham and D. K. Pradhan. The de Bruijn multiprocessor network: A versatile parallel processing and sorting network for VLSI. *IEEE Trans. on Computers*, 38(4), 1989.

[69] N. Schmitt and R. F. Farwell. *Understanding Automation Systems.* Texas Instruments, Dallas, TX, 1984.

[70] A. Segall. Distributed network protocols. *IEEE Trans. Information Theory,* IT-20(1):23–35, 1983.

[71] M. B. Sharma, S. S. Iyengar, and M. K. Mandyam. An efficient distributed depth-first-search algorithm. *Info Processing Letters,* 32:183–186, 1989.

[72] R. A. Short. The attainment of reliable digital systems through the use of redundancy: A survey. *IEEE Computer Group News,* 2:2–17, 1968.

[73] D. P. Siewiorek and R. S. Swarz. *The Theory and Practice of reliable System Design.* Digital Press, 1982.

[74] R. G. Smith. *A Framework for Distributed Problem Solving.* University of Michigan Press, Ann Arbor, 1981.

[75] R. G. Smith. Frameworks for cooperation in distributed problem solving. *IEEE Trans. SMC,* pages 61–70, Jan. 1981.

[76] A. Tanenbaum. *Distributed Operating Systems.* Prentice Hall, Engelwood Cliffs, NJ, 1995.

[77] R. R. Tenney and N. R. Sandell. Detection with distributed sensors. *IEEE Trans. Aerospace Electronic Systems,* AES-17(4):98–101, 1981.

[78] J. N. Tsitsiklis and M. Athans. On the complexity of decentralized decision making and detection problems. *IEEE Trans. Automatic Control,* AC-30(5):440–446, 1985.

[79] John von Neumann. Probabilistic logics and the synthesis of reliable organisms from unreliable components. In C. E. Shannon and J. McCarthy, editors, *Automata Studies,* pages 43–98. Princeton Univeristy Press, Princeton, NJ, 1956.

[80] R. Wesson, F. Hayes-Roth, J. W. Burge, C. Stasz, and C. A. Sunshine. Network structures for distributed situation assessment. *IEEE Trans. Systems, Man, Cybernetics,* SMC 11(1):5–23, 1981.

[81] Y. Yemini. Distributed sensor networks: An attempt to define the issues. In *Proc. Distributed Sensor Network Workshop,* pages 53–66, 1978. Carnegie Mellon University, Pittsburgh, PA.

[82] Y. F. Zheng. Integration of multiple sensors into a robotic system and its performance evaluation. *IEEE Trans. Robotics, Automation,* 5(5):658–669, 1989.

[83] S. S. Iyengar and Weian Deng. An Efficient Edge Detection Algorithm using Relaxation Labeling Technique. *Journal of Pattern Recognition*, 28(4) 1995.

[84] J. Shen and S. Castin. Further Results on DRF Method for Edge Detection. *9th Intl. Conf. on Pattern Recognition*, Rome, 1988.

[85] L. G. Roberts. Machine Perception of Three Dimensional Solids. *Optical and Electro-Optical Information Processing*, MIT Press, Cambridge, Massachusets, 1965.

[86] I. Sobel. Camera Models and Machine Perception. *Stanford AI Memo 121*, Dept. of Computer Science, Stanford University, 1970.

[87] J. Canny. Computational Approach to Edge Detection. *IEEE Trans. on Pattern Analysis and Machine Intelligence*, 8(6):669–714, 1986.

[88] S. S. Iyengar and Guna Seetharaman. Multisensory Perception of Three Dimensional Scene. To be published, 1994.

[89] Z. Dong. Artificial Intelligence in Optimal Design and Manufacturing. Volume III, Prentice Hall Series on Envrionment and Intellignet Manufacturing System. Prentice Hall, Englewood Cliffs, New Jersey.

[90] R. R. Brooks and S. S. Iyengar. Multi-Dimensional Sensor Integration: Theory and Practice. Submitted for Publication, 1994.

[91] L. Prasad, S. S. Iyengar, R. L. Kashyap and R. L. Rao. Fault-Tolerant Sensor Integration Using Multiresolution Decomposition *J. Physical Review E*, 49(4), pp 3452–3461, 1994.

[92] L. Prasad, S. S. Iyengar. A General computational framework for distributed sensing and fault-tolerent sensor integration. *IEEE Trans. System, Man, Cybernetics*, 25(4), pp 643–650.

[93] Y. Meyer. Wavelets: Algorithms and applications. *SIAM*, Philadelphia, 1993.

Index

About the Authors

Dr. S. S. Iyengar - Chairman of the Computer Science Department and Professor of Computer Science at Louisiana State University. He has been involved with research in high-performance algorithms and data structures since receiving his PhD in 1974, and has directed over 22 PhD dissertations at LSU. He has served as a principal investigator on research projects supported by the Office of Naval Research, the National Aeronautics and Space Administration, the National Science Foundation, California Institute of Technology's Jet Propulsion Laboratory, the Department of Navy-NORDA, the Department of Energy, LEQFS - Board of Regents and the U.S. Army Office.

His publications include several books and over 200 publications (including 105 archival journal papers) in areas of high-performance parallel and distributed algorithms and data structure for image processing and pattern recognition, autonomous navigation, and distributed sensor networks. He was a visiting professor at Jet Propulsion Laboratory, Oak Ridge National Laboratory, and Indian Institute of Science. He is a series editor for *Neuro Computing of Complex Systems*, and area editor for the *Journal of Computer Science and Information*. He has served as guest editor for the *IEEE Transactions on Software Engineering*, the *IEEE Transactions on System, Man, and Cybernetics* and the *IEEE Transactions on Knowledge and Data Engineering*. He was awarded the Williams Evans Fellowship from the University of Otago, New Zealand in 1991. He is a Fellow of the IEEE for his research contributions in the area of Algorithms and Data Strucures in Image Processing.

Dr. L. Prasad - is a member of the research staff at the Los Alamos National Laboratory. His interest is in the design and analysis of algorithms for nonlinear systems. He has published a large number of papers in these areas. He graduated with a Ph.d in Computer Science from Louisiana State University, in April 1995.

Hla Min - Obtained M.S. degree in Electrical Engineering from Rice University and is nearing completion of his Ph.D from Louisiana State University. His area of interest is algorithms for computer vision.